ANIMAL TISSUE CULTURE

ANIMAL TISSUE CULTURE

A. WILSON ARUNI

Research Assistant Professor
Division of Microbiology and Molecular Genetics
School of Medicine
Loma Linda University
California, USA

P. RAMADASS

Professor and Head (Retired)
Department of Animal Biotechnology
Madras Veterinary College
Chennai, Tamil Nadu

MJP Publishers

Cataloguing-in-Publication Data

Aruni, A. Wilson (1967-   ).
    Animal Tissue Culture / by A. Wilson Aruni
and P. Ramadass. –
Chennai : MJP Publishers, 2011
    xviii, 192 p.; 24 cm.
    Includes Glossary, References and Index.
    ISBN 978-81-8094-056-9 (pb.)
    1. Tissue Culture, Animal
I. Ramadass, P. II Title.
571.538 dc 22 ARU MJP 091

ISBN 978-81-8094-056-9
© Publishers, 2011
All rights reserved
Printed and bound in India

MJP PUBLISHERS
47, Nallathambi Street
Triplicane
Chennai 600 005

Publisher : J.C. Pillai
Managing Editor : C. Sajeesh Kumar
Marketing Manager : S.Y. Sekar
Project Editor : P. Parvath Radha
Acquisitions Editor : C. Janarthanan
Editorial Team : B. Ramalakshmi, N. Thilagavathi
Lissy John, M. Gnanasoundari,
CIP Data : Prof. K. Hariharan, Librarian
RKM Vivekananda College, Chennai.

This book has been published in good faith that the work of the author is original. All efforts have been taken to make the material error-free. However, the author and publisher disclaim responsibility for any inadvertent errors.

FOREWORD

Tissue culture has been one of the most important techniques that have paved the way for the unprecedented growth in the field of cell and molecular biology. This book on *Animal Tissue Culture* incorporates fundamental knowledge and recent developments in the field of *in vitro* culturing and manipulation of eukaryotic cells. The chapters encompass a comprehensive presentation of the subject matter which should satisfy the reader. Though the book targets the undergraduate and post-graduate curriculums, chapters on transgenic animals and stem cells presents a most consistent version of up-to-date knowledge that expands the scope of this book. The figures and photographic plates, in a simplified manner, give a clear understanding of the practical utility and protocols related to *in vitro* assays, their principles and techniques. Chapters on scale-up techniques explain the principles and industrial use of cell culture related techniques. Thus this book will be an asset to the academic, research and industrial communities. It depicts the practical experience, commitment and invaluable expertise of the authors in completing this project.

Hansel Fletcher

Hansel M. Fletcher, Ph.D.
Professor & Vice Chair
Graduate Program Coordinator

A Seventh-day Adventist Institution

DEPARTMENT OF BASIC SCIENCES, DIVISION OF MICROBIOLOGY AND MOLECULAR GENETICS
11021 Campus Street, Room 101, Loma Linda, California 92350
(909) 558-4480 · *fax* (909) 558-4035 · www.llu.edu

PREFACE

Cell and tissue culture has been one of the first and foremost techniques paving way for recent cutting-edge technologies such as vaccinology, monoclonal antibody production, therapeutic cloning, stem cell technology, etc. It has played a substantial role in the developments of health care and prophylactics industries, thus serving the mankind. It has made the dream of producing cost-effective prophylactics, diagnostics and therapeutics come true and affordable.

In the recent past, with the explosion of knowledge in the field of biotechnology, intensive research is being carried out, where undergraduate and post-graduate courses are being offered in this field. Even though more emphasis is being given to theory, a dearth of practical knowledge is lacking due to paucity of established tissue culture facilities. This technique has become a common affordable one in the recent past with blooming educational technology in the field of Microbiology and Biotechnology, where institutions resort to setting up tissue culture labs and pilot studies. But as of now, a comprehensive book covering the theory as well as practical approach is lacking. Though a few books on animal tissue culture authored by foreign authors were resorted to earlier, adaptation of these techniques indigenously were less documented and published.

This book *Animal Tissue Culture* satiates the quest for such lacunae by bringing the theory-based practical knowledge on the protocols and standard operating techniques involving animal tissue culture. The book covers a formal introduction to the subject and brings about the basics in understanding the concept of tissue culture traversing through the state-of-the-art technologies such as stem cell, hybridoma technology, transgenics, cloning and scaling-up technologies. The photographic plates showing results of different tissue-culture-based techniques stands unique and indigenous and many are from our own work. From the students' standpoint, this book will impart the up-to-date knowledge and confidence to execute tissue culture-related work with ease. The chapters covered are based on the postgraduate and undergraduate curriculum followed by majority of the universities.

We have an extensive experience in teaching and research in the field of tissue culture and have also authored two related books namely *Animal Biotechnology: Recent Concepts and Developments* and *Practical Biotechnology*.

Though this book contains protocols and subjects in common to tissue culture techniques, many of them have been standardized by our work in the laboratory. However we do not claim them to be our original contribution because they include some compilations from different methods and from various sources. A number of standard textbooks, reviews and journals were consulted to make this book authenticated and to maintain standards. We express our apologies for any mistake inadvertently committed during the preparation of this book. Any suggestion to improve the book in future editions is welcome.

A. Wilson Aruni

P. Ramadass

CONTENTS

1

INTRODUCTION TO
ANIMAL CELL CULTURE

INTRODUCTION

Cell culture has become one of the major tools used in the life sciences today. **Tissue Culture** is the general term for the removal of cells, tissues, or organs from an animal or plant and their subsequent placement into an artificial environment conducive to growth. This environment usually consists of a suitable glass or plastic culture vessel containing a liquid or semi-solid medium that supplies the nutrients essential for survival and growth. The culture of whole organs or intact organ fragments with the intent of studying their continued function or development is called **organ culture**. When cells are removed from the organ fragments prior to, or during cultivation, thus disrupting their normal relationships with neighbouring cells, it is called **cell culture**.

Although animal cell culture was first successfully undertaken by Ross Harrison in 1907, it was not until the late 1940s to early 1950s that several developments occurred that made cell culture widely available as a tool for scientists. First, there was the development of antibiotics that made it easier to avoid many of the contamination problems that plagued earlier cell culture attempts. Second, there was the development of the techniques such as the use of trypsin to remove cells from culture vessels, necessary to obtain continuously growing cell lines (such as HeLa cells). Third, using these cell lines, scientists were able to develop standardized, chemically defined culture media that made it far easier to grow cells. These developments paved the way for many more scientists to use cell, tissue and organ culture in their research.

PROCESS OF CULTURING CELLS

PRIMARY CULTURE

When cells are surgically removed from an organism and placed into a suitable culture environment, they will attach, divide and grow. This is called a **primary culture**. There are two basic methods for doing this. First, for **explant cultures**, small pieces of tissue are attached to a glass or treated plastic culture vessel and bathed in culture medium. After few days, individual cells will move from the tissue explant out onto the culture vessel surface or substrate where they will begin to divide and grow. The second, more widely used method speeds up this process by adding digesting (proteolytic) enzymes such as trypsin or collagenase to the tissue fragments to dissolve the cement holding the cells together. This creates a suspension of single cells that are then placed into culture vessels containing culture medium and allowed to grow and divide. This method is called **enzymatic dissociation**.

SUBCULTURING

When the cells in the primary culture vessel have grown and filled up all of the available culture substrate, they must be **subcultured** to give them room for continued growth. This is usually done by removing them as gently as possible from the substrate with enzymes. These are similar to the enzymes used in obtaining the primary culture and are used to break the protein bonds attaching the cells to the substrate. Some cell lines can be harvested by gently scraping the cells off the bottom of the culture vessel. Once released, the cell suspension can then be subdivided and placed into new culture vessels. Once a surplus of cells is available, they can be treated with suitable cryoprotective agents such as dimethylsulphoxide (DMSO) or glycerol, carefully frozen and then stored at cryogenic temperatures (below −130°C) until they are needed.

CELL CULTURE SYSTEMS

Two basic culture systems are used for growing cells. These are based primarily upon the ability of the cells to either grow attached to a glass or treated plastic substrate (**monolayer culture sytems**) or floating free in the culture medium (**suspension culture systems**). Monolayer cultures are usually grown in tissue culture treated dishes, T-flasks, roller bottles, or multiple well plates, the choice being based on the number of cells needed, the nature of the culture environment, cost and personal preference. Suspension cultures are usually grown in either of the following:

1. Magnetically rotated spinner flasks or shaken Erlenmeyer flasks where the cells are kept actively suspended in the medium;
2. Stationary culture vessels such as T-flasks and bottles where, although the cells are not kept agitated, they are unable to attach firmly to the substrate. Many cell lines,

especially those derived from normal tissues, are considered to be **anchorage-dependent**, that is, they can only grow when attached to a suitable substrate.

Some cell lines that are no longer considered normal (frequently designated as **transformed cells**) are frequently able to grow either attached to a substrate or floating free in suspension; they are **anchorage-independent**. In addition, some normal cells such as those found in the blood do not normally attach to substrates and always grow in suspension.

TYPES OF CELLS

Cultured cells are usually described based on their morphology (shape and appearance) or their functional characteristics.

There are three basic morphologies:

1. **Epithelial-like** cells that are attached to a substrate and appear flattened and polygonal in shape.
2. **Lymphoblast-like** cells that do not attach normally to a substrate but remain in suspension with a spherical shape.
3. **Fibroblast-like** cells that are attached to a substrate and appear elongated and bipolar, frequently forming swirls in heavy cultures.

It is important to remember that the culture conditions play an important role in determining shape and that many cell cultures are capable of exhibiting multiple morphologies. Using cell fusion techniques, it is also possible to obtain hybrid cells by fusing cells from two different parents. These may exhibit characteristics of either parent or both parents. This technique was used in 1975 to create cells capable of producing custom-tailored monoclonal antibodies. These hybrid cells (called **hybridomas**) are formed by fusing two different but related cells. The first is a spleen-derived lymphocyte that is capable of producing the desired antibody. The second is a rapidly dividing myeloma cell (a type of cancer cell) that has the machinery for making antibodies but is not programmed to produce any antibody. The resulting hybridomas can produce large quantities of the desired antibody. These antibodies, called **monoclonal antibodies** due to their purity, have many important clinical, diagnostic, and industrial applications.

FUNCTIONAL CHARACTERISTICS OF CULTURED CELLS

The characteristics of cultured cells result from both their origin (liver, heart, etc.) and how well they adapt to the culture conditions. Biochemical markers can be used to determine if cells are still carrying on specialized functions that they perform *in vivo* (e.g., liver cells secreting albumin). Morphological or ultrastructural markers can also be examined (e.g., beating heart cells). Frequently, these characteristics are either lost or changed as a result of being placed in an artificial environment. Some cell lines will eventually stop dividing and show signs of aging. These lines are called **finite**. Other cell lines become immortal; these can continue to divide indefinitely and are called **continuous** cell lines. When a "normal" finite cell line becomes immortal, it has undergone a fundamental irreversible change or "transformation". This can occur spontaneously or be brought about intentionally using drugs, radiation or viruses. **Transformed cells** are usually easier and faster growing, may often have extra or abnormal chromosomes and frequently can be grown in suspension. Cells that have the normal number of chromosomes are called **diploid** cells, those that have other than the normal number are **aneuploid**. If the cells form tumours when they are injected into animals, they are considered to be **neoplastically transformed**.

PROBLEMS FACED BY CULTURED CELLS

AVOIDING CONTAMINATION

Cell culture contamination is of two main types: chemical and biological. Chemical contamination is the most difficult to detect since it is caused by agents such as endotoxins, plasticizers, metal ions or traces of chemical disinfectants, which are invisible. Biological contaminants in the form of fast growing yeast, bacteria and fungi usually have visible effects on the culture (changes in medium turbidity or pH) and thus are easier to detect (especially if antibiotics are omitted from the culture medium). However, two other forms of biological contamination, mycoplasmas and viruses, are not easy to detect visually and usually require special detection methods.

There are two major requirements to avoiding contamination:

1. Proper training in and use of good aseptic technique on the part of the cell culturist.
2. Properly designed, maintained and sterilized equipment, plasticware, glassware, and media.

The careful and selective (limited) use of antibiotics designed for use in tissue culture can also help avoid culture loss due to biological contamination.

GOOD ENVIRONMENT FOR THE CELLS TO GROW *IN VITRO*

To cell culturists, a "GOOD" environment is one that does more than allowing the cells to survive in culture. Usually, it means an environment that, at the very least, allows cells to increase in number by undergoing cell division (mitosis). Even better, when conditions are just right, some cultured cells will express their "happiness" with their environment by carrying out important *in vivo* physiological or biochemical functions, such as muscle contraction or the secretion of hormones and enzymes. To provide this environment, it is important to provide the cells with the appropriate temperature, a good substrate for attachment, and the proper culture medium.

Temperature is usually set at the same point as the body temperature of the host from which the cells were obtained. With cold-blooded vertebrates, a temperature range of 18°C to 25°C is suitable; most mammalian cells require 36°C to 37°C. This temperature range is usually maintained by use of carefully calibrated, and frequently checked, incubators. Anchorage-dependent cells also require a good substrate for attachment and growth. Glass and specially treated plastics (to make the normally hydrophobic plastic surface hydrophilic or wettable) are the most commonly used substrates. However, **attachment factors**, such as collagen, gelatin, fibronectin and laminin, can be used as substrate coatings to improve growth and function of normal cells derived from brain, blood vessels, kidney, liver, skin, etc. Often normal anchorage-dependent cells will also function better if they are grown on a permeable or porous surface. This allows them to polarize (have a top and bottom through which things can enter and leave the cell) as they do in the body. Many specialized cells can only be truly normal in function when grown on a porous substrate in serum-free medium with the appropriate mixture of growth and attachment factors. Cells can also be grown in suspension on beads made from glass, plastic, polyacrylamide and cross-linked dextran molecules. This technique has been used to enable anchorage-dependent cells to be grown in suspension culture systems and is increasingly important for the manufacture of cell-based biologicals. The culture medium is the most important and complex factor to control in making cells functionally good. Besides meeting the basic nutritional requirement of the cells, the culture medium should also have any necessary growth factors, regulate the pH and osmolality, and provide essential gases (O_2 and CO_2). The 'food' portion of the culture medium consists of amino acids, vitamins, minerals, and carbohydrates. These allow the cells to build new proteins and other components essential for growth and function as well as for providing the energy necessary for metabolism.

The growth factors and hormones help regulate and control the cells' growth rate and functional characteristics. Instead of being added directly to the medium, they are often added in an undefined manner by adding 5 to 20% of various animal sera to the medium. Unfortunately, the types and concentration of these factors in serum vary considerably from batch to batch. This often results in problems controlling growth and function. When growing normal functional cells, sera are often replaced by specific growth factors.

The medium also controls the pH range of the culture and buffers the cells from abrupt changes in pH. Usually a CO_2–bicarbonate based buffer or an organic buffer, such as HEPES, is used to help keep the medium pH in the range 7.0–7.4 depending on the type of cell being cultured. When using a CO_2–bicarbonate buffer, it is necessary to regulate the amount of CO_2 dissolved in the medium. This is usually done using an incubator with CO_2 controls set to provide an atmosphere with between 2% and 10% CO_2 (for Earle's salt-based buffers). However, some media use a CO_2–bicarbonate buffer (for Hanks' salt-based buffers) that requires no additional CO_2, but it must be used in a sealed vessel (not dishes or plates).

Finally, the osmolality (osmotic pressure) of the culture medium is important since it helps regulate the flow of substances in and out of the cell. It is controlled by the addition or subtraction of salt in the culture medium. Evaporation of culture media from open culture vessels (dishes, etc.) will rapidly increase the osmolality.

BASIC ENVIRONMENTAL
REQUIREMENTS FOR GOOD CELLS

- Controlled temperature
- Good substrate for cell attachment
- Appropriate medium and incubator that maintains the correct pH and osmolality. Suspension and microcarrier cultures can be grown in glass spinner vessels. CHO-K1 cells growing on a microcarrier bead resulting in stressed, damaged or dead cells. For open culture systems, incubators with high humidity levels to reduce evaporation are essential.

JUDGING OF CULTURED CELLS

Evaluating the general health of a culture is usually based on four important cell characteristics:

- morphology
- growth rate
- plating efficiency
- expression of special functions

These same characteristics are also widely used in evaluating experimental results. The **morphology** or cell shape is the easiest to determine but is often the least useful. While changes in morphology are frequently observed in cultures, it is often difficult to relate these observations to the condition that caused them. It is also a very difficult characteristic to quantify or to measure precisely. Often, the first sign that something is wrong with a culture occurs when the cells are microscopically examined and poor or unusual patterns of cell attachment or growth are observed. When problems are suspected, staining the culture vessels with crystal violet or other simple histological stains may show growth patterns indicating a problem.

Cell counting and other methods for estimating cell number, on the other hand, allow the determination of the **growth rate**, which is sensitive to major changes in the culture environment. This allows the design of experiments to determine which set of conditions (culture media, substrate, serum, plasticware) is better for the cells, i.e., the conditions producing the best growth rate. These same or similar techniques can also be used to measure cell survival or death and are often used for *in vitro* cytotoxicity assays.

Plating efficiency is a testing method where small numbers of cells (20 to 200) are placed in a culture vessel and the number of colonies they form is measured.

The percentage of cells forming colonies is a measure of survival, while the colony size is a measure of growth rate. This testing method is similar in application to growth rate analysis but is more sensitive to small variations in culture conditions. The final characteristic, the **expression of specialized functions**, is usually the most difficult to observe and measure. Usually biochemical or immunological assays and tests are used. While cultured cells may grow very well in suboptimal conditions, highly specialized functions usually require near-perfect culture conditions and are often quickly lost when cells are placed in culture.

GENERAL USES OF CELL CULTURE

Cell culture has become one of the major tools used in cell and molecular biology. Some of the important areas where cell culture is currently playing a major role are the following.

Model systems Cell cultures provide a good model system for studying

1. basic cell biology and biochemistry,
2. the interactions between disease-causing agents and cells,
3. the effects of drugs on cells,
4. the process and triggers for aging, and
5. nutritional studies.

Toxicity testing Cultured cells are widely used alone or in conjunction with animal tests to study the effects of new drugs, cosmetics and chemicals on survival and growth in a wide variety of cell types. Especially important are liver- and kidney-derived cell cultures.

Cancer research Since both normal cells and cancer cells can be grown in culture, the basic differences between them can be closely studied. In addition, it is possible, by the use of chemicals, viruses and radiation, to convert normal cultured cells to cancer-causing cells. Thus, the mechanisms that cause the change can be studied. Cancer cells also serve as a test system to determine suitable drugs and methods for selectively destroying various types of cancer.

Virology One of the earliest and major uses of cell culture is the replication of viruses in cell cultures (in place of animals) for use in vaccine production. Cell cultures are also widely used in the clinical detection and isolation of viruses, as well as basic research into how they grow and infect organisms.

Cell-based manufacturing While cultured cells can be used to produce many important products, three areas are generating the most interest.

First is the large-scale production of viruses for use in vaccine production. These include vaccines for polio, rabies, chickenpox, hepatitis B and measles.

Second, is the large-scale production of cells that have been genetically engineered to produce proteins that have medicinal or commercial value. These include monoclonal antibodies, insulin, hormones, etc.

Third, is the use of cells as replacement tissues and organs. Artificial skin for use in treating burns and ulcers is the first commercially available product. However, research and testing is underway on artificial organs such as pancreas, liver and kidney. A potential supply of replacement cells and tissues may come out of work currently being done with both embryonic and adult stem cells. These are cells that have the potential to differentiate into a variety of different cell types. It is hoped that learning how to control the development of these cells may offer new treatment approaches for a wide variety of medical conditions.

Genetic counselling   Amniocentesis, a diagnostic technique that enables doctors to remove and culture foetal cells from pregnant women, has given doctors an important tool for the early diagnosis of foetal disorders. These cells can then be examined for abnormalities in their chromosomes and genes using karyotyping, chromosome painting and other molecular techniques.

Genetic engineering   The ability to transfect or reprogram cultured cells with new genetic material (DNA and genes) has provided a major tool to molecular biologists wishing to study the cellular effects of the expression of theses genes (new proteins). These techniques can also be used to produce these new proteins in large quantity in cultured cells for further study. Insect cells are widely used as miniature cell factories to express substantial quantities of proteins that they manufacture after being infected with genetically engineered baculoviruses.

Gene therapy   The ability to genetically engineer cells has also led to their use for gene therapy.

Cells can be removed from a patient lacking a functional gene and the missing or damaged gene can then be replaced. The cells can be grown for a while in culture and then replaced into the patient. An alternative approach is to place the missing gene into a viral vector and then "infect" the patient with the virus in the hope that the missing gene will then be expressed in the patient's cells.

Drug screening and development   Cell-based assays have become increasingly important for the pharmaceutical industry, not just for cytotoxicity testing but also for high throughput screening of compounds that may have potential use as drugs. Originally, these cell culture tests were done in 96 well plates.

REVIEW QUESTIONS

1. Write short notes on the following
    i. cell culture systems
    ii. judging of cultured cells
    iii. types of cells

2. Write in detail the general uses of cell culture.

2

INITIATION OF CELL CULTURE

SAFETY ASPECTS OF CELL CULTURE

RISK ASSESSMENT

The main aim of risk assessment is to prevent injury, protect property and avoid harm to individuals and the environment. Consequently risk assessments must be undertaken prior to starting any activity. The assessment consists of two elements:

1. Identifying and evaluating the risks.
2. Defining ways of minimizing or avoiding the risk.

For animal cell culture, the level of risk is dependent upon the cell line to be used and is based on whether the cell line is likely to cause harm to humans. The different classifications are given below:

Low risk	• Non-human/non-primate continuous cell lines and some well characterized human diploid lines of finite lifespan (e.g., MRC-5).
Medium risk	• Poorly characterized mammalian cell lines.
High risk	• Cell lines derived from human/primate tissue or blood.
	• Cell lines with endogenous pathogens (the precise categorization is dependent upon the pathogen).
	• Cell lines used following experimental infection where the categorization is dependent upon the infecting agent.

A culture collection recommends a minimum containment level required for a given cell line based upon its risk assessment. For most cell lines the appropriate level of containment is Category 2. However, this may need to be increased to Category 3 depending upon the type of manipulations to be carried out and whether large culture volumes are envisaged. (The containment levels are classified as I, II, III based on the risk factor to the researcher).

Containment is the most important means of reducing risk. Other less significant measures include restricting the movement of staff and equipment into and out of laboratories. Good laboratory practice and good bench techniques such as ensuring that work areas are uncluttered, reagents are correctly labelled and stored, are also important for reducing risk and making the laboratory a safe environment in which to work.

DISINFECTION

Methods designed for the disinfection/decontamination of culture waste, work surfaces and equipment represent important means for minimizing the risk of harm.

The major disinfectants fall into four groups and their relative merits can be summarized as follows:

Hypochlorites These are good general-purpose disinfectants.

- Active against viruses
- Corrosive against metals and therefore should not be used on metal surfaces, e.g., centrifuges
- Readily inactivated by organic matter and therefore should be made fresh daily
- Should be used at 1000 ppm for general use surface disinfection, 2500 ppm in discard waste pots for washing pipettes, and 10,000 ppm for tissue culture waste and spillage

Note: When fumigating a cabinet or room using formaldehyde all the hypochlorites must first be removed, as the two chemicals react together to produce carcinogenic products.

Phenolics These are not active against viruses. They remain active in the presence of organic matter.

Alcohol (e.g., ethanol, isopropanol)

- Effective concentrations are 70% for ethanol and 60–70% for isopropanol.
- Their mode of activity is by dehydration and fixation.
- Effective against bacteria. Ethanol is effective against most viruses but not against nonenveloped viruses.
- Isopropanol is not effective against viruses.

Aldehydes (e.g., glutaraldehyde, formaldehyde)

- Aldehydes are irritants and their use should be limited due to problems of sensitization.
- Glutaraldehyde may be used in situations where the use of hypochlorites is not suitable, e.g., cleaning of centrifuge bowls or materials constructed of stainless steel that may be attacked or corroded by using hypochlorite solutions.

WASTE DISPOSAL

Given below is a list of ways in which tissue culture waste can be decontaminated and disposed of safely. One of the most important aspects of the management of all laboratory-generated waste is to dispose of waste regularly and not to allow the amounts to build-up. Different forms of waste require different treatment.

- **Tissue culture waste** (culture medium) Inactivate overnight in a solution of hypochlorite (10,000 ppm) prior to disposal to drain with an excess of water.
- **Contaminated pipettes** should be placed in hypochlorite solution (2500 ppm) overnight before disposal by autoclaving and incineration.
- **Solid waste** such as flasks, centrifuge tubes, contaminated gloves, tissues, etc. should be placed inside heavy-duty sacks for contaminated waste and autoclaved prior to incineration. These bags are available from Bibby Sterilin and Greiner.
- If at all possible, waste should be incinerated rather than autoclaved.

DESIGN AND EQUIPMENT FOR THE CELL CULTURE LABORATORY

LABORATORY DESIGN

One of the most under-rated aspects of tissue culture is the need to design the facility to ensure that good-quality material is produced in a safe and efficient manner. Most tissue culture is undertaken in laboratories that have been adapted for the purpose and in conditions that are not ideal. However, as long as a few basic guidelines are adopted, this should not compromise the work.

There are several aspects to the design of good tissue culture facilities. Ideally work should be conducted in a single use facility which, if at all possible, should be separated into an area reserved for handling newly received material (quarantine area) and an area for material which is known to be free of contaminants (main tissue culture facility). If this is not possible, work should be separated by time with all manipulations on clean material being completed prior to manipulations involving the "quarantine material". Different incubators should also be designated. In addition, the work surfaces should be thoroughly cleaned between activities. All new material should be handled as "quarantine material" until it has been shown to be free of contaminants such as bacteria, fungi and particularly mycoplasma. Conducting tissue culture in a shared facility requires considerable planning and it is essential that a good technique is used throughout to minimize the risk of contamination occurring.

However, the precise category required is dependent upon the cell line and the nature of the work proposed. The guidelines make recommendations regarding the laboratory environment including lighting, heating, the type of work surfaces and flooring and provision of handwashing facilities. In addition it is recommended that laboratories should be run at air pressures that are negative to corridors to contain any risks within the laboratory.

Microbiological Safety Cabinets

A microbiological safety cabinet is probably the most important piece of equipment since, when operated correctly, it will provide a clean working environment for the product, whilst protecting the operator from aerosols. In these cabinets operator and/or product protection is provided through the use of HEPA (high efficiency particulate air) filters. The level of containment provided varies according to the class of cabinet used. Cabinets may be ducted to atmosphere or re-circulated through a second HEPA filter before passing to atmosphere.

In most cases a class II cabinet is adequate for animal cell culture. However each study must be assessed for its hazard risk and it is possible that additional factors, such as a known virus infection or an uncertain provenance, may require a higher level of containment.

Centrifuges

Centrifuges are used routinely in tissue culture as part of the subculture routine for most cell lines and for the preparation of cells for cryopreservation. By their very nature, centrifuges produce aerosols and thus it is necessary to minimize this risk. This can be achieved by purchasing models that have sealed buckets. Ideally the centrifuge should have a clear lid so that the condition of the load can be observed without opening the lid. This will reduce the risk of the operator being exposed to hazardous material if a centrifuge tube has broken during centrifugation. Care should always be taken not to over-fill the tubes and to balance them carefully. These simple steps will reduce the risk of aerosols being generated. The centrifuge should be situated where it can be easily accessed for cleaning and maintenance. Centrifuges should be checked frequently for signs of corrosion.

Incubators

Cell cultures require a strictly controlled environment in which to grow. Specialist incubators are used routinely to provide the correct growth conditions, such as temperature, degree of humidity and CO_2 levels in a controlled and stable manner. Generally they can be set to run at temperatures in the range 28°C (for insect cell lines) to 37°C (for mammalian cell lines) and set to provide CO_2 at the required level (e.g., 5–10%). Some incubators also have the facility to control the O_2 levels. Copper-coated incubators are also now available. These are reported to reduce the risk of microbial contamination within the incubator due to the microbial inhibitory activity of copper. The inclusion of water bath treatment fluid in the incubator water trays will also reduce the risk of bacterial and fungal growth in the water trays. However, there is no substitute for regular cleaning.

WORK SURFACES AND FLOORING

In order to maintain a clean working environment the laboratory surfaces including benchtops, walls and flooring should be smooth and easy to clean. They should also be waterproof and resistant to a variety of chemicals (such as acids, alkalis, solvents and disinfectants). In areas used for the storage of materials in liquid nitrogen, the floors should be resistant to cracking if any liquid nitrogen is spilt. In addition, the floors and walls should be continuous with a covered skirting area to make cleaning easier and reduce the potential for dust to accumulate. Windows should be sealed. Work surfaces should be positioned at a comfortable working height.

PLASTICWARE AND CONSUMABLES

Almost every type of cell culture vessel, together with support consumables such as tubes and pipettes, are commercially available as single-use, sterile-packed, plasticware. Suppliers include Sigma-Aldrich, Nunc, Greiner, Bibby Sterilin and Corning. The use of such plasticware is more cost-effective than recycling glassware, enables a higher level of quality assurance and removes the need for validation of cleaning and sterilization procedures. Plastic tissue culture flasks are usually treated to provide a hydrophilic surface to facilitate attachment of anchorage-dependent cells.

CARE AND MAINTENANCE OF LABORATORY AREAS

In order to maintain a clean and safe working environment, tidiness and cleanliness are the key. Obviously all spills should be dealt with immediately. The cleaning of all work surfaces both inside and outside of the microbiological safety cabinet, the floors and all other pieces of equipment, e.g., centrifuges should also be routinely undertaken. Humidified incubators are a particular area for concern due to the potential for fungal and bacterial growth in the water trays. This will create a contamination risk that can be avoided only by regular cleaning of the incubator. All major pieces of equipment should be regularly maintained and serviced by qualified engineers.

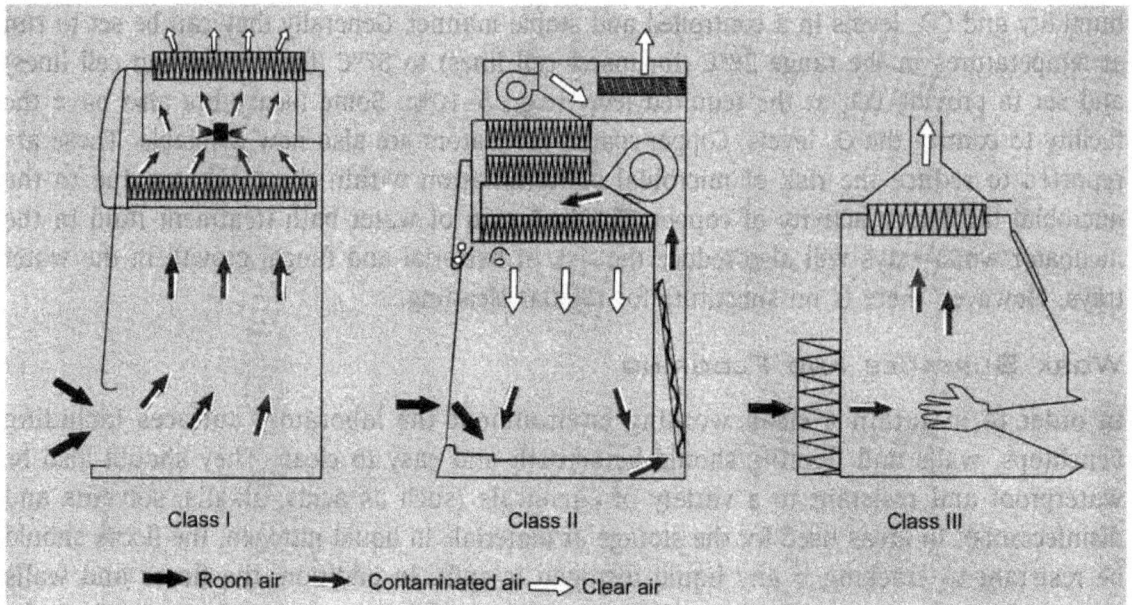

Class I   Class II   Class III

➡ Room air   ⇨ Contaminated air ⟹ Clear air

Figure 2.1 Various levels of biosafety cabinets (From *Fundamental Techniques to Cell Culture, A Laboratory Handbook,* Sigma-Aldrich co.)

PROCESS OF INITIATION OF CELL CULTURE

Cell culture is the process by which prokaryotic, eukaryotic or plant cells are grown under controlled conditions. In practice the term "cell culture" has come to refer to the culturing of cells derived from multicellular eukaryotes, especially animal cells. The historical development and methods of cell culture are closely interrelated to those of tissue culture and organ culture.

Animal cell culture became a routine laboratory technique in the 1950s, but the concept of maintaining live cell lines separated from their original tissue source was discovered in the 19th century.

History

The 19th-century English physiologist Sydney Ringer developed salt solutions containing the chlorides of sodium, potassium, calcium and magnesium suitable for maintaining the beating of an isolated animal heart outside of the body. In 1885 Wilhelm Roux removed a portion of the medullary plate of an embryonic chicken and maintained it in a warm saline solution for several days, establishing the principle of tissue culture. Ross Granville Harrison, working at Johns Hopkins Medical School and then at Yale University, published results of his experiments from 1907–1910, establishing the methodology of tissue culture.

Cell culture techniques were advanced significantly in the 1940s and 1950s to support research in virology. Growing viruses in cell cultures allowed preparation of purified viruses for the manufacture of vaccines. The Salk polio vaccine was one of the first products mass-produced using cell culture techniques. This vaccine was made possible by the cell culture research of John Franklin Enders, Thomas Huckle Weller, and Frederick Chapman Robbins, who were awarded the Nobel Prize for their discovery of a method of growing the virus in monkey kidney cell cultures.

ISOLATION OF CELLS

Cells can be isolated from tissues for *ex vivo* culture in several ways. Cells can be easily purified from blood; however only the white cells are capable of growth in culture. Mononuclear cells can be released from soft tissues by **enzymatic digestion** with enzymes such as collagenase, trypsin, or pronase, which break down the extracellular matrix. Alternatively, pieces of tissue can be placed in growth media, and the cells that grow out are available for culture. This method is known as **explant culture.**

Cells that are cultured directly from a subject are known as **primary cells.** With the exception of some derived from tumours, most primary cell cultures have limited lifespan. After a certain number of population doublings, cells undergo the process of senescence and stop dividing, while generally retaining viability.

An established or immortalized **cell line** has acquired the ability to proliferate indefinitely either through random mutation or deliberate modification, such as artificial expression of the telomerase gene. There are numerous well-established cell lines representative of particular cell types.

MAINTAINING CELLS IN CULTURE

Cells are grown and maintained at an appropriate temperature and gas mixture (typically, 37°C, 5% CO_2) in a cell incubator. Culture conditions vary widely for each cell type, and variation of conditions for a particular cell type can result in different phenotypes being expressed.

Aside from temperature and gas mixture, the most commonly varied factor in culture systems is the growth medium. Recipes for growth media can vary in pH, glucose concentration, growth factors, and the presence of other nutrient components. The growth factors used to supplement media are often derived from animal blood, such as calf serum. These blood-derived ingredients pose the potential for contamination of derived pharmaceutical products with viruses or prions. Current practice is to minimize or eliminate the use of these ingredients where possible.

Some cells naturally live without attaching to a surface, such as cells that exist in the bloodstream. Others require a surface, such as most cells derived from solid tissues. Cells grown unattached to a surface are referred to as **suspension cultures**. Other **adherent culture** cells can be grown on tissue culture plastic, which may be coated with extracellular matrix components to increase its adhesion properties and provide other signals needed for growth. **Organotypic cultures** involve growing cells in a three-dimensional environment as opposed to two-dimensional culture dishes. This 3D culture system is biochemically and physiologically more similar to *in vivo* tissue, but is technically challenging to maintain.

MANIPULATION OF CULTURED CELLS

As cells generally continue to divide in culture, they generally grow to fill the available area or volume. This can generate several issues:

- Nutrient depletion in the growth media
- Accumulation of apoptotic/necrotic (dead) cells.
- Cell-to-cell contact can stimulate cell cycle arrest, causing cells to stop dividing, known as contact inhibition.
- Cell-to-cell contact can stimulate promiscuous and unwanted cellular differentiation.

These issues can be dealt with using tissue culture methods that rely on **sterile technique**. These methods aim to avoid contamination with bacteria or yeast that will compete with mammalian cells for nutrients and/or cause cell infection and cell death. Manipulations are typically carried out in a biosafety hood or laminar flow cabinet to exclude contaminating microorganisms. Antibiotics can also be added to the growth media. Amongst the common manipulations carried out on culture cells are media changes, passaging cells, and transfecting cells.

MEDIA CHANGES

The purpose of media changes is to replenish nutrients and avoid the build-up of potentially harmful metabolic by-products and dead cells. In the case of suspension cultures, cells can be separated from the media by centrifugation and resuspended in fresh media. In the case of adherent cultures, the media can be removed directly by aspiration and replaced.

PASSAGING CELLS

Passaging or splitting cells involves transferring a small number of cells into a new vessel. Cells can be cultured for a longer time if they are split regularly, as it avoids the senescence associated with prolonged high cell density. Suspension cultures are easily passaged with a small amount of culture containing a few cells diluted in a larger volume of fresh media. For adherent cultures, cells first need to be detached; this is commonly done with a trypsin–EDTA mixture, however other enzyme mixes are now available for this purpose. A small number of detached cells can then be used to seed a new culture.

TRANSFECTION AND TRANSDUCTION

Another common method for manipulating cells involves the introduction of foreign DNA by transfection. This is often performed to cause cells to express a protein of interest. More recently, the transfection of RNAi constructs have been realized as a convenient mechanism for suppressing the expression of a particular gene/protein.

DNA can also be inserted into cells using viruses, in methods referred to as transduction, infection or transformation. Viruses, as parasitic agents, are well-suited to introducing DNA into cells, as this is a part of their normal course of reproduction.

REVIEW QUESTIONS

1. Write short notes on
    i. Disinfectants used in cell culture laboratory
    ii. Transfection and transduction
    iii. Biosafety cabinets

2. Explain in detail design and equipments for a cell culture laboratory.

3

GROWTH OF CELLS IN CULTURE

Cell culture is the process by which prokaryotic or eukaryotic cells are grown under controlled conditions. In practice, the term "cell culture" refers to the culturing of cells derived from multicellular eukaryotes, especially animal cells.

METHODS OF CULTURES

There are three main methods of initiating a culture, which are listed below.

1. *Organ culture*   This implies that the architectural characteristic of the tissue *in vivo* is retained, at least in part, in the culture. Tissue is cultured at the liquid–gas interface (on raft, grid, gel, etc.), which favours the retention of a spherical or three-dimensional shape.

2. *Primary explant culture*   In this method, a fragment of tissue is placed on a glass or plastic–liquid interface, where after attachment, migration is promoted in the plane of the solid substrate.

3. *Cell culture*   It implies that the tissue, or outgrowth from the primary explant, is dispersed (mechanically or enzymatically) into a cell supension, which may then be cultured as an adherent monolayer on a solid substrate or as a suspension in the culture medium.

The properties of the three types of cultures are listed in Table 3.1.

Table 3.1 Properties of different types of cultures

Category	Organ culture	Explant culture	Cell culture
Source	Embryonic organs, adult tissue fragment	Tissue fragment	Disaggregated tissue, primary culture, propagated cell line.
Effort	High	Moderate	Low
Characterization	Histology	Cytology and markers	Biochemical, molecular, immunological, and cytological assays.
Histology	Informative	Difficult	Not applicable
Biochemical differentiation	Possible	Heterogeneous	Lost, but may be reinduced
Propagation	Not possible	Possible from outgrowth	Standard procedure
Replicate sampling, reproducibility, homogeneity	High intersample variation	High intersample variation	Low intersample variation
Quantitation	Difficult	Difficult	Easy; many techniques are available.

For establishing a cell culture, the cells are obtained directly from animal tissue (primary explant) or from a culture collection. Isolation directly from explant offers a means of culturing cells that will be more genotypically close to their *in vivo* state, but the isolation process is cumbersome and demanding compared to establishing a culture from a stock that could be obtained from standard culture collections. The latter is the recommended route for obtaining cells that have been well-characterized in terms of their growth, origin, and genetic traits.

A **primary culture** is established when the cells taken directly from animal tissue are added to growth medium. Cells which are capable of proliferation will survive and grow. Primary culture is the first step in the selection process that may ultimately give rise to a relatively uniform **cell line**. The formation of the cell line from a primary culture implies

- an increase in the total number of those cultured cells for several generations,
- the ultimate predominance of that cell or its lineage, and
- a degree of uniformity in the cell population.

In primary explantation, selection occcurs by virtue of the cells' capacity to migrate from the explant. But with dispersed cells, only those cells that both survive the disaggregation technique (viz., enzymatic or mechanical) and adhere to the substrate or survive in suspension will form the basis of a primary culture. If primary cell is maintained for more than a few hours, a further selection step will occur—cells that are capable of proliferation and increase, some cell types may survive but not increase, and yet others will be unable to survive. Hence the relative proportion of each cell type will change and will continue to do so until, in the case of monolayer cultures, all the available culture substrate is occupied. This is called **confluence**, a phenomenon associated with cell–cell interaction. Population changes and adaptive modifications within the cell are occurring continuously throughout the culture, making it difficult to select a period when the culture may be regarded as homogeneous or stable.

After confluence is reached (i.e., all available growth area is utilized and the cells make close contact with one another), cells will stop dividing due to **contact inhibition**. When cells stop growing in culture, new cultures can be established by inoculating some of the cells to fresh medium. This is called **subculturing** or **passaging**. It is important to subculture within a day or so of the maximum cell density to ensure continued growth in a new culture. Cells will lose their viability if they are left for too long before subculture. In order to obtain a reasonable growth rate, cells should be inoculated at a density of 10^4–10^5 cells/ml.

For cells grown in suspension, **subculture** involves dilution of the high-density culture with fresh medium. Dilutions from 1 : 2 to 1 : 10 v/v will be suitable. The subculture of anchorage-dependent cells involves detachment of the cells from the growth surface (substratum) of one culture flask and reinoculation of the cells to fresh medium contained in a new flask. The cells are detached from their anchor by a process of **trypsinization**. The proteolytic enzyme, trypsin, is used to break down the proteins that bind the cells to the culture surface. The trypsin is added to the washed cells in the culture flask for a short period which is long enough to dislodge the cells from the substratum but not too long to damage the cells. The action of trypsin is stopped by the addition of serum-supplemented medium and centrifugation to remove the cells.

After first subculturing or passage, the primary culture becomes known as a **cell line** and may be propagated and subcultured several times. Cultures are given a **passage number** which indicates the number of subcultures performed since the cells were obtained or isolated. With each successive passaging, the component of population with the ability to proliferate more will be diluted out. Selected-out population will be able to withstand trauma of trypsinization and transfer. Although some selection and phenotypic drift will continue, by the third passage, the culture becomes more stable and is typified by a rather hardy, rapidly proliferating cell. In the presence of serum and without specific selection conditions, cells like fibroblasts frequently overgrow the culture. This overgrowth represents one of the major challenges of tissue culture since its inception, especially for maintenance of specialized cell lines.

Normal cells can divide only for a limited number of times; hence cell lines derived from normal tissue will die out after a fixed number of population doublings. This is a genetically determined event involving several genes and is known as **senescence**. It is thought to be determined, in part, by the inability of terminal sequences of DNA in the telomeres to replicate at each cell division. The result is a progressive shortening of the telomeres until, finally, the cell is unable to divide further. Exceptions to this are stem cells, germ cells and transformed cells, which often have an enzyme telomerase capable of replicating the terminal sequence of DNA in the telomere and extending the lifespan of the cells.

Some cell lines may give rise to **continuous cell lines** or **established cell lines**. They are immortal by genetic modification. Genetic variation often involves the deletion or mutation of *p53* gene, which would normally arrest cell cycle progression. An alteration in a culture that gives rise to a continuous cell line is commonly called *in vitro* **transformation**. It can be spontaneous or chemically or virally induced. **Immortalization** means the acquisition of an infinite lifespan and **transformation** implies an additional alteration in growth characteristics (like anchorage-dependence, loss of contact inhibition, etc.). Transformation often correlates with **tumorigenicity**. The vast majority of established cell lines are derived from tumours (e.g., HeLa) or from cells transformed *in vitro,* although some of the very earliest lines were established from normal embryonic tissue (e.g., 3T3, CHO). There are also lines that have been widely used, such as WI-38, which are from normal human tissues and have a limited lifespan *in vitro*.

Continuous cell lines are aneuploid and often have a chromosome number between the diploid and tetraploid values. There is also considerable variation in chromosome number and constitution among cells in population (heteroploidy). Many normal cells will not give rise to continuous cell lines. Normal human fibroblasts remain euploid throughout their lifespan, and will stop dividing, at crisis, although they remain viable up to 18 months thereafter. Human glia and chick fibroblasts behave similarly. Epidermal cells on the other hand have shown gradually increasing lifespan with improvements in culture techniques and eventually give rise to continuous cell lines.

These cell lines have been the workhorses of cell culture, from their use in studying the control of the cell cycle to vaccine production and large-scale industrial production of recombinant proteins. Not surprisingly, after many decades of growth in many laboratories, they are relatively tough (i.e., resistant to temporary lapses in good cell culture technique) and have altered from their original phenotype. Thus, cells having the same designation carried in different laboratories may vary considerably in their properties.

More recently, cell lines have been developed with the aim of maintaining a normal phenotype combined with the ability to grow the cell, or its precursor, indefinitely in culture. This can be accomplished using conditional transformation or by establishing the cell line from stem cell or precursor cells, which can then be induced to differentiate into a terminally differentiated cell type in culture. These lines are generally more challenging to handle *in vitro*.

When cells are selected from a culture by cloning or by some other methods, the subline is known as a **cell strain**. The detailed characterization is implied. **Histotypic culture or histoculture** means a high density or tissue-like culture of one cell type whereas organotypic culture implies the presence of more than one cell-type interaction us they might in the organ of origin. The **organotypic culture** has given new prospects for the study of cell interaction among the discrete, defined population of homogeneous and potentially, genetically and phenotypically defined cells.

PHASES OF GROWTH OF CELLS IN CULTURE

In order to analyse the growth characteristics of a particular cell type or cell line, a **growth curve** can be established from which one can obtain a **population doubling time, a lag time, and a saturation density**. A growth curve (Figure 3.1) generally follows different stages which are as follows:

a. Lag phase It is the time taken for the cell to recover from subculture, attach, and spread.

b. Log phase In this phase the cell number begins to increase exponentially.

c. Plateau phase or stationary phase In this phase the culture becomes confluent and there is no increase in cell concentration. In the plateau phase, cell growth is limited by many factors like nutrient depletion, accumulation of toxic by-products and lack of surface for adherence.

d. Decline phase In this phase the growth rate slows down or stops.

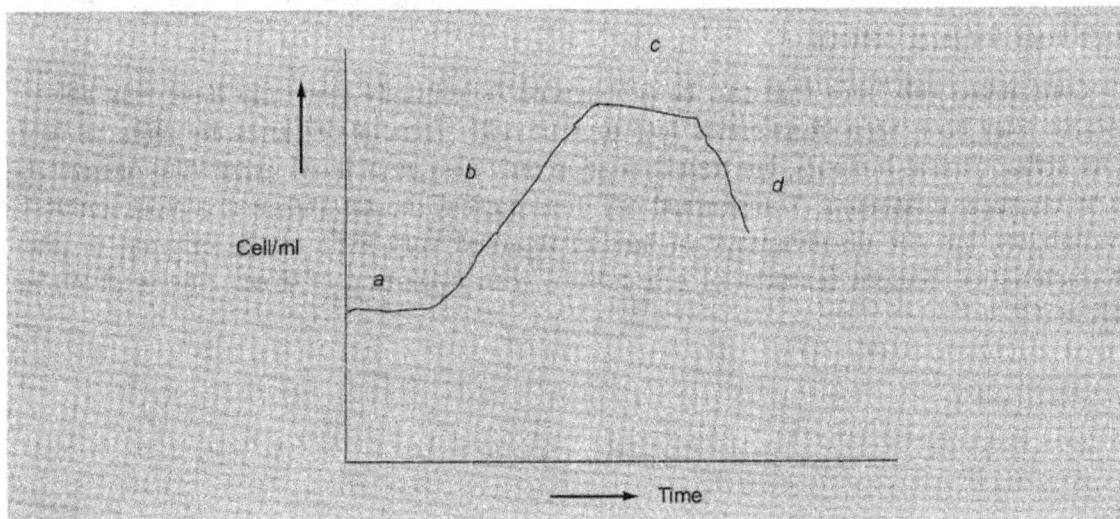

Figure 3.1 Cell growth in culture (a—lag phase, b—log phase, c—stationary phase, d—decline phase)

While growth curves provide much information, they are too time- and labour-consuming to be used where changes in cell number are to be used as a screen. **Growth**, or increase in total cell number over time, is a good measure of a biological response because it is so broadly defined and influenced by many different factors, including mitogens, changes in nutrient level, transport, membrane integrity, attachment factors, and so forth. An increase in cell number is also a frequently used method of assessing the effect of hormones, nutrients, and so forth on a specific cell type.

MAIN TYPES OF CELL CULTURE

PRIMARY CULTURES

Primary cultures are derived directly from excised, normal animal tissue and cultured either as an explant culture or following dissociation into a single cell suspension by enzyme digestion. Such cultures are initially heterogeneous but later become dominated by fibroblasts. The preparation of primary cultures is labour-intensive and they can be maintained *in vitro* only for a limited period of time. During their relatively limited lifespan, primary cells usually retain many of the differentiated characteristics of the cell *in vivo*.

CONTINUOUS CULTURES

Continuous cultures are comprised of a single cell type that can be serially propagated in culture either for a limited number of cell divisions (approximately thirty) or otherwise indefinitely. Cell lines of a finite life are usually diploid and maintain some degree of differentiation. The fact that such cell lines senesce after approximately thirty cycles of division means it is essential to establish a system of Master and Working banks in order to maintain such lines for long periods.

Continuous cell lines that can be propagated indefinitely generally have this ability because they have been transformed into tumour cells. Tumour cell lines are often derived from actual clinical tumours, but transformation may also be induced using viral oncogenes or by chemical treatments. Transformed cell lines present the advantage of almost limitless availability, but the disadvantage of having retained very little of the original *in vivo* characteristics. Salient features of cell culture with evolution of a cell line is given in Figure 3.2.

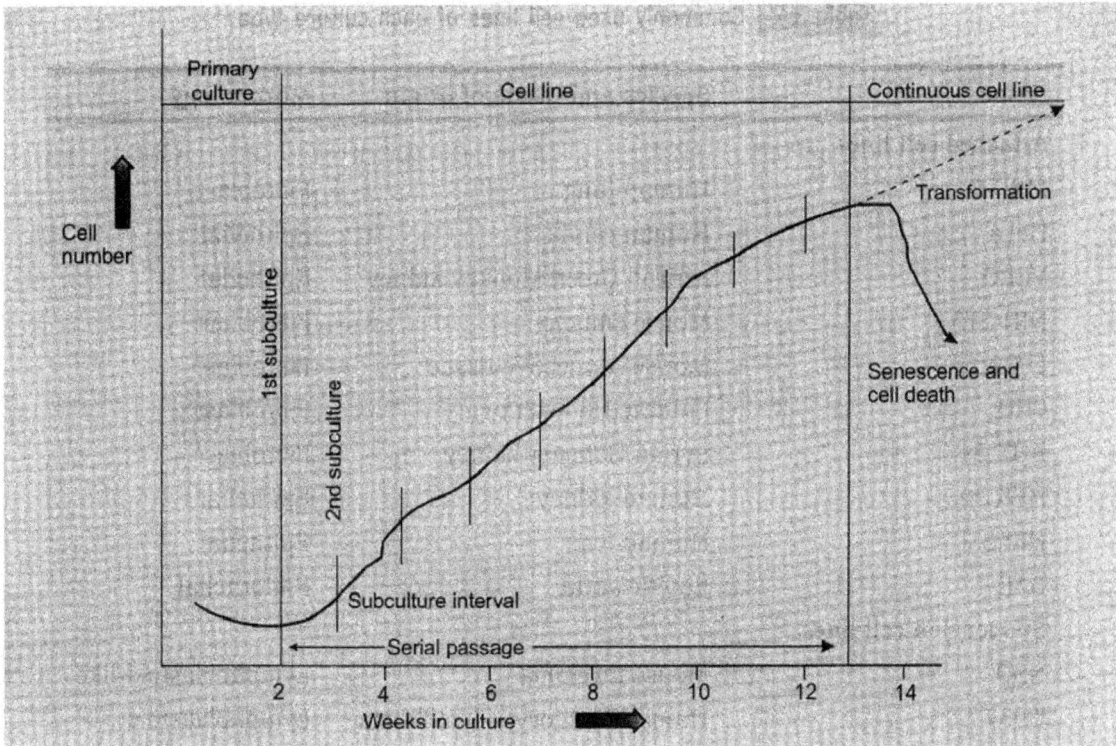

Figure 3.2 Salient features of cell culture with evolution of a cell line

CULTURE MORPHOLOGY

Morphologically cell cultures take one of the two forms, growing either in suspension (as single cells or small free-floating clumps) or as a monolayer that is attached to the tissue culture flask (Table 3.2). The form taken by a cell line reflects the tissue from which it was derived, e.g., cell lines derived from blood (leukaemia, lymphoma) tend to grow in suspension whereas cells derived from solid tissue (lungs, kidney) tend to grow as monolayers. Attached cell lines can be classified as endothelial such as BAE-1, epithelial such as HeLa, neuronal such as SH-SY5Y or fibroblasts such as MRC-5 and their morphology reflect the area within the tissue of origin.

There are some instances when cell cultures may grow as semi-adherent, cells e.g., B95-8 where there appears to be a mixed population of attached and suspension cells. For these cell lines it is essential that both cell types are subcultured to maintain the heterogeneous nature of the culture.

Table 3.2 Commonly used cell lines of each culture type

Name	Species and tissue of origin	Morphology
Attached cell lines		
MRC-5	Human lung	Fibroblast
HeLa	Human cervix	Epithelial
VERO	African Green Monkey kidney	Epithelial
NIH 3T3	Mouse embryo	Fibroblast
L929	Mouse connective tissue	Fibroblast
CHO	Chinese Hamster ovary	Fibroblast
BHK-21	Syrian Hamster kidney	Fibroblast
HEK 293	Human kidney	Epithelial
HEPG 2	Human liver	Epithelial
BAE-1	Bovine aorta	Endothelial
Suspension cell lines		
NSO	Mouse myeloma	Lymphoblastoid-like
U937	Human Hystiocytic lymphoma	Lymphoblastoid
Namalwa	Human lymphoma	Lymphoblastoid
HL60	Human leukaemia	Lymphoblastoid-like
WEHI 231	Mouse B-cell lymphoma	Lymphoblastoid
YAC 1	Mouse lymphoma	Lymphoblastoid
U 266B1	Human myeloma	Lymphoblastoid
SH-SY5Y	Human neuroblastoma	Neuroblast

SUBCULTURING/PASSAGING OF CELL LINES

Most animal cell lines and primary cultures grow as a single thickness cell layer or sheet attached to a plastic or glass substrate. Once the available substrate surface is covered by cells (a **confluent** culture), growth slows down and then ceases. Thus, in order to keep the cells healthy and actively growing, it is necessary to subculture them at regular intervals. Usually, this subcultivation process involves breaking the bonds or cellular 'glue' that attaches the cells to the substrate and to each other by using proteolytic enzymes such as trypsin, dispase, or collagenase. Occasionally, these enzymes or dissociating agents are combined with divalent cation chelators such as EDTA (binds calcium and magnesium ions). The loosened cells are then removed from the culture vessel, counted, diluted and subdivided into new vessels. Cells then reattach, begin to grow and divide, and, after a suitable incubation period (depending on the initial inoculum size, growth conditions and cell line), again reach saturation or confluency. At this point, the subcultivation cycle can be repeated. The following protocol covers the basic techniques that are suitable for subculturing many cell lines.

PROTOCOL FOR SUBCULTURING

1. Sterile flask of actively growing cells that are 80 to 90% confluent.
2. *Cell culture medium*   This should contain all of the additives (foetal bovine serum, glutamine, etc.) required by the above cell line.
3. *Calcium- and magnesium-free phosphate-buffered saline CMF-PBS* (10 mL)   This simple salt solution is used to maintain proper pH and osmotic balance while the cells are being washed to remove protease inhibitors that are found in most animal sera.
4. *0.1% trypsin solution*   Trypsin is normally used in concentrations ranging from 0.05% to 0.25%. Working concentrations are usually determined by using the lowest trypsin concentration that can remove the cells from the substrate and give a single cell suspension in a relatively short time (5 to 10 minutes). Trypsin solutions are often supplemented with other enzymes (collagenase) or chelating agents (EDTA) to improve its performance.
5. 15 mL disposable screw-capped centrifuge tubes.
6. Appropriate culture vessels.

Procedure

Examination   It is important to examine your cultures daily and always prior to subcultivation. Using an inverted phase-contrast microscope (100 to 200×), quickly check the general appearance of your culture. Look for signs of microbial contamination. Many cells round up during mitosis, forming very refractile (bright) spheres that may float free of the surface when the culture is disturbed. Dead cells often round up and become detached but are usually not bright or refractile.

Cell harvesting   This step removes the cells from the plastic substrate and breaks cell-to-cell bonds as gently as possible.

When using enzymatic dissociation:

   i. the old medium is removed and discarded;

   ii. the cell monolayer is gently rinsed;

   iii. the enzyme solution is added and the culture incubated until the cells are released.

Do not forget to examine the culture vessel with the unaided eye to look for small fungal colonies that may be floating at the media–air interface (especially near the vessel neck) and thus not visible through the microscope.

A phosphate-buffered saline is used for both rinsing and trypsinization since it maintains a physiological pH without requiring a closed system (required by buffers based on Hanks' saline) or gassing with carbon dioxide (required by buffers based on Earle's saline). Calcium and magnesium are omitted because these play a role in cell attachment.

i. Using a sterile pipette, remove and discard the culture medium. All materials and solutions exposed to cells must be disposed of properly. Medium can be left in the pipettes if they are placed in disinfectant.

ii. For a T-75 flask, wash the cell monolayer by adding 5 to 10 mL of CMF-PBS to the flask and then slowly rock it back and forth to remove all traces of foetal bovine serum. Remove and discard the wash solution. Failure to remove traces of foetal bovine serum is frequently responsible for failure of the trypsin solution to remove the cells from the vessel. Proportionally reduce or increase the volumes used in this protocol for smaller or larger culture vessels.

iii. Add 5 mL of the trypsin solution (in CMF-PBS) to the flask and place the flask back in an incubator at 37°C to increase the activity of the enzyme solution. (Prewarming of the enzyme solution to 37°C will decrease the required exposure period.)

iv. Check the progress of the enzyme treatment every few minutes with an inverted phase-contrast microscope. Once most of the cells have rounded up, gently tap the side of the flask to detach them from the plastic surface. Then add 5 mL of growth medium to the cell suspension and, using a 10 mL pipette, vigorously wash any remaining cells from the bottom of the culture vessel. At this point a quick check on the inverted microscope should show that the cell suspension consists of at least 95% single cells. If this is not the case, more vigorous pipetting may be necessary.

v. Collect the suspended cells in a 15 mL centrifuge tube and place on ice. Some dissociating agents should be removed at this point by centrifugation to prevent carry-over which can cause poor cell attachment or toxicity. However, the trypsin in the cell suspension will be inactivated by the serum and does not absolutely need to be removed. If removal is desired, spin the cell suspension at 100 ×g for 5 minutes. Then remove the trypsin containing medium and replace with fresh medium.

Cell counting   To determine growth rates or set-up cultures at known concentrations, it is necessary to count the cell suspension. Haemocytometers or electronic cell counting devices can be used. The haemocytometer has the added advantages of being less expensive and allowing cell viability determinations to be made during counting.

i. Vortex the cell suspension and remove a 0.5 mL sample and place in a tube for counting. To this add 1mL of the vital stain trypan blue (0.04%). Mix well by vortexing, withdraw a 20 μL sample with a wide-tip pipettor and carefully load a clean haemocytometer.

ii. Do a viable cell count and calculate the number of viable cells/mL and the total cell number.

Storing cells on ice will slow cell metabolism. This will improve cell viability and reduce cell clumping. Frequently, instead of counting the cells in the suspension, the suspension is split among a number of culture vessels. For example, a 1:2 split would divide the cell suspension of one vessel into two new vessels of equivalent surface area. This is a quick and easy method for the routine maintenance of cell lines.

Two CMF-PBS washes and/or rinsing with trypsin can be used for more difficult-to-remove cells. For difficult-to-break-up cell clumps, try holding the pipette tip tight against and perpendicular to the side of the flask and then forcibly expel its contents. This will create a strong shearing force that should break up cell clumps.

Plating After making the appropriate dilutions, add the correct amount of cells to each culture vessel. Then add fresh medium to bring the culture vessel to its recommended working volume (Table 3.3). Label all vessels accurately; write on the sides of flasks and around the outer edge of the dish tops so as not to interfere with microscopic observation.

Table 3.3 Typical cell yields and recommended medium volumes for flasks and dishes

Type of vessel	Average cell yield	Recommended medium volume (mL)	Maximum working volume (mL)
Flasks			
25 cm^2	2.5×10^6	5–7.5	10
75 cm^2	7.5×10^6	15–22.5	60
150 cm^2	1.5×10^7	30–45	210
162 cm^2	1.6×10^7	32–48	250
175 cm^2	1.75×10^7	35–52.5	370
225 cm^2	2.25×10^7	45–67.5	
Dishes			
35 mm	8.0×10^5	1.6–2.4	NA
60 mm	2.1×10^6	4.2–6.3	NA
100 mm	5.5×10^6	10–15	NA
150 mm	1.48×10^7	30–45	NA
245 mm	5.0×10^7	100–150	NA

Incubation   Most mammalian cell cultures do best at a temperature between 35° and 37°C. In addition to maintaining constant temperature, some incubators also maintain high humidity levels and CO_2 concentrations. The high humidity cuts down evaporation losses in open systems such as Petri dishes and microplates that would otherwise result in hypertonic culture medium and stressed cells. The elevated CO_2 concentrations (usually 5% to 10%, depending on bicarbonate concentrations in the medium) help maintain the proper pH (7.4 ± 0.2) when used with the correct bicarbonate buffer system. In order for this type of buffer system to work, it is necessary to allow gas exchange by using unsealed dishes and plates or flasks with gas-permeable (vented) caps.

i. Leave caps on flasks slightly loosened or use vented caps on the flasks to allow gas exchange as well as for extra protection against spillage and contamination and place on a shelf in a 37°C, humidified CO_2 incubator. It is recommended to use 0.2 to 0.3 mL of medium for every square centimetre of growth area.

ii. Examine cultures daily and change medium as needed.

Note   It is assumed that there is an average yield of 1×10^5 cm^2 from a 100% confluent culture. The maximum working volume is the amount a flask can hold in the horizontal position when filled to the neck.

REVIEW QUESTIONS

1. What are the different methods of culture?
2. Explain different phases of growth of cells in culture with a growth curve.
3. What are the steps involved in subculturing of cells?

4

IMPORTANCE OF ASEPTIC TECHNIQUES IN CELL CULTURE

The techniques by which cell culture procedures are executed without introducing contaminating microorganisms from the environment are called aseptic or sterile techniques. Contamination by microorganisms remains a major problem in tissue culture. The major contaminants are bacteria, mycoplasma, yeast and fungal spores and the cross-contamination with other cell lines. Aseptic techniques aim to exclude this contamination by establishing a strict code of practice and ensuring that everyone using the facility adheres to it. Slow growth of animal cells and richness of nutrients in the medium make the culture particularly vulnerable to the fast growth of potential microbial contaminants.

ELEMENTS OF ASEPTIC ENVIRONMENT

Correct aseptic techniques should provide a barrier between the microorganisms in the environment outside the culture and the pure, uncontaminated culture within the flask or dish. Hence, all materials that will come into direct contact with the culture must be sterile and the manipulations must be designed such that there is no direct link between the culture and its non-sterile surroundings. A strict aseptic technique should be followed for the same and a few of it is detailed below.

A WELL-DESIGNED LABORATORY

A conveniently designed laboratory is a major prerequisite to maintain aseptic conditions in a cell culture laboratory. All new materials should be handled as "quarantine materials" until it has been shown to be free of contaminants such as bacteria, fungi and particularly *Mycoplasma*.

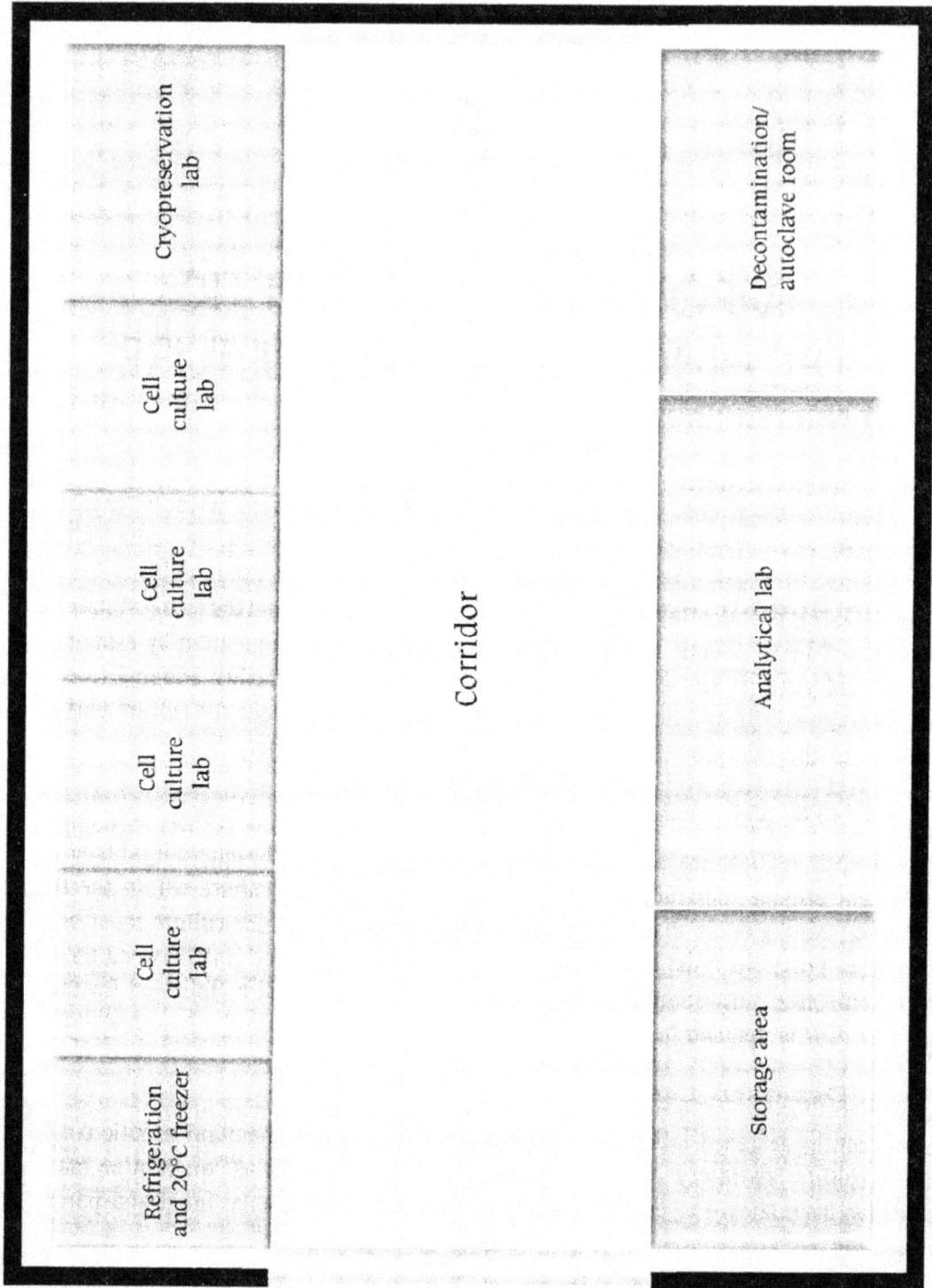

Figure 4.1 The cell culture laboratory and supporting facilities

For most cell lines the laboratory should be designated to at least Category 2 based on international biosafety considerations. Ideally, the space allocated for the tissue culture laboratory should be one dedicated to tissue culture functions exclusively, in order to minimize the introduction of potential contaminants. Traffic in and out of the culture room (space) and talking in the space are to be discouraged. All tasks that do not need to be performed in the culture room (e.g., which do not require a sterile environment) should be performed elsewhere. People not actively engaged in doing cell culture should leave the room. Minimized entry and exit may be achieved, for example, by having a refrigerator and freezer in the culture room or an airlock "entry room" so that there is no need to enter and exit the culture room during the course of an experiment to obtain reagents necessary for the culture work. If space allows, an airlock can help to ensure a "clean" tissue culture room. If it is not possible to have a separate room for the cell culture equipment, select a corner of the laboratory that is farthest away from doors and other heavily trafficked areas may be selected. All the culture equipment is placed together in this area of the room and any equipment not needed for cell culture is transferred to another area of the laboratory. This area should then be cleaned and maintained as described.

The need to minimize the potential for contamination requires that the room be kept under positive pressure with high-efficiency particulate air (HEPA-filtered air) flowing through it. The floors should be smooth and untextured. If a vinyl floor covering is used, it should be a continuous unseamed sheet. False ceilings are also a potential source of contamination and should not be used with a positive pressure air flow. If possible, a solid ceiling should be constructed. Minimally, new ceiling tiles should be installed every few years (if you see stained or damaged tiles or mould growing on or between the tiles, it is time!) and the space above well-cleaned, with any leaks, from the outside or from condensation, fixed immediately. Plumbing and all other "bulkhead" fittings and hardware which pass through the wall or ceiling should be well-sealed.

Laminar Flow Cabinet

There are two types of laminar flow hoods, vertical and horizontal. The vertical hood, also known as a biological safety cabinet (Figure 4.2), is best for working with hazardous organisms since the aerosols that are generated in the hood are filtered out before they are released into the surrounding environment. Horizontal hoods (Figure 4.3) are designed such that the air flows directly at the operator hence they are not useful for working with hazardous organisms but are the best protection for your cultures. Both types of hoods have continuous displacement of air that passes through a HEPA filter that removes particulates from the air. In a vertical hood, the filtered air blows down from the top of the cabinet; in a horizontal hood, the filtered air blows out at the operator in a horizontal fashion.

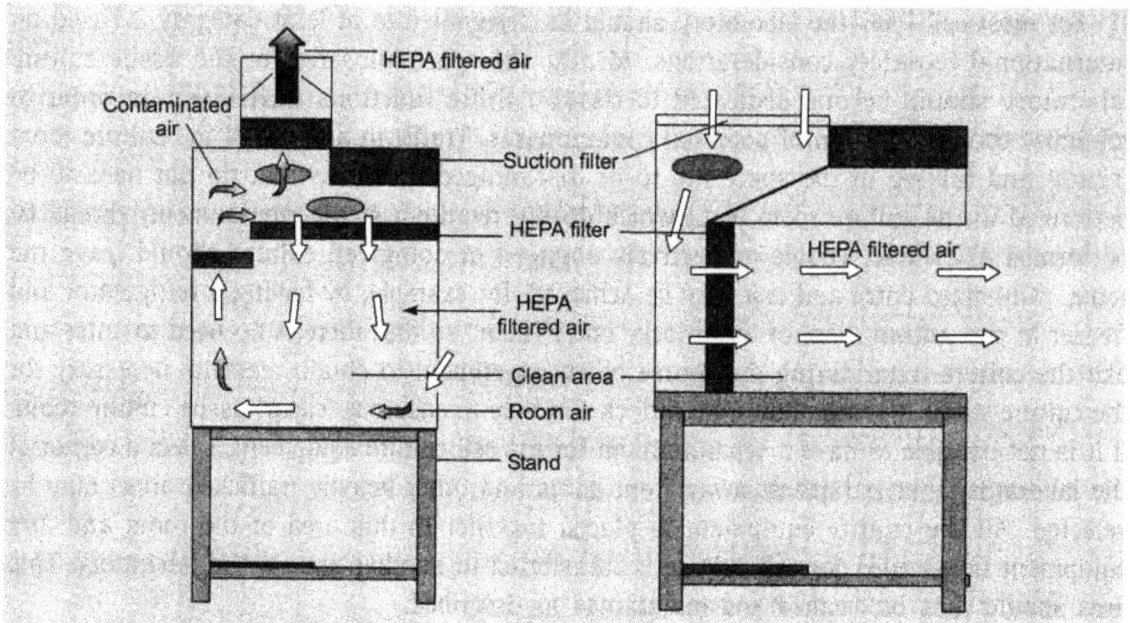

Figure 4.2 Biological safety hood

Figure 4.3 Horizontal laminar flow

- Laminar flow cabinet should be placed in an area that is free from air currents from door, windows, etc. The area should have no traffic and no equipment that generates air current should be positioned so that air currents do not compromise the functioning of the hood.

- The activity should be restricted to tissue culture and no animals and microbiological cultures should be brought to the tissue culture area.

- The cabinet should be kept clean free of dust and should not contain equipment other than that collected with tissue culture.

- Non-sterile activities such as sample processing, staining or extraction should not be carried out in the cabinet.

- Standard operating procedures should be followed for laminar flow cabinet.

- Environmental monitoring with tryptose soya broth agar settle plates inside the cabinet for a minimum of four hours should be a good indicator of how clean a cabinet is. There should be no growth of bacteria or fungi on such plates.

- In most cases a class II cabinet is adequate for animal cell culture. However each study must be assessed for its hazard risk and it is possible that additional factors, such as a known virus infection or an uncertain provenance, may require a higher level of containment.

WORK SURFACE

It is essential to keep the work surface clean and tidy. The following rules should be followed.

1. Start with a completely clear surface.
2. Swab the surface liberally with 70% alcohol.
3. Bring on to the surface only those items required for a particular procedure.
4. Remove everything that is not required, and swab the surface down between the procedures.
5. Arrange your work area conveniently with an easy access to all items without having to reach over one to get at another (especially over an open bottle or flask).
6. Horizontal laminar flow chamber is more forgiving but should work in a clear space with no obstruction between the central work area and HEPA filter.
7. Work within a range of vision (e.g., insert a pipette in a bulb or pipetting aid with the tip of the pipette pointing away from you so that it is in your line of sight continuously and not hidden by your arm).
8. Mop up any spillage immediately and swab the area with 70% alcohol.
9. Use sterile wrapped pipettes and discard them after use into a biohazard waste container.
10. Check that the wrapping of the sterile pipette is not broken or damaged.
11. Inspect the vessels to be used:
    a. T-flask—must be free from visible contamination or breakage, or lack container identification. The plastic covering the flask must be intact.
    b. Bottles—check for cracks, expiry dates.
    c. Spinner flasks—check for cracks, expiry dates, and proper assembly.
12. Discard any biohazardous or contaminated material immediately.
13. Never perform mouth pipetting. Pipettor must be used.
14. When handling sterile containers with caps or lids, place the cap on its side if it must be laid on the work surface.
15. Make sure not to touch the tip of the pipette to the rim of any flask or sterile bottle.
16. Remove everything when the work is finished and swab the work surface down again.

REAGENTS AND MEDIA

Reagents and media obtained from commercial suppliers will have undergone strict quality control to ensure that they are sterile. But the outside surface of the bottle may not be sterile and the wrapping outside the hood should be removed. Unwrapped bottles should be swabbed in 70% alcohol that come from the refrigerator or from a water bath.

Tissue culture media used are often supplemented with antibiotics. Antibiotics do not eliminate problems of gross contamination which result from poor sterile technique or antibiotic-resistant mutants. Autoclaving renders pipettes, glassware, and solutions sterile.

Nutrient medium cannot be autoclaved. The compounds in nutrient medium are destroyed by the heat of autoclaving. Medium must therefore be sterilized by passing it through a sterile filter small enough in pore size to hold back bacteria and mycoplasmas (Millipore Sterivex-GS 0.22 μm disposable filter units).

CULTURES

Cultures imported from another laboratory carry a high risk because they may have been contaminated either at the source or in transit. Important cell lines should always be quarantined. They should be handled separately from the rest of our stocks and kept free of antibiotics until they are shown to be uncontaminated. Antibiotics should not be used routinely as they may only suppress and not eliminate some contaminations.

PERSONAL HYGIENE

Personal protective equipment usually called PPE should be a strict practice in cell culture labs. Footware, lab coats, gloves, eye and face protecting aids and respirators are the common PPE in cell culture lab. Washing hands is a good practice to reduce microbial contamination. Surgicals may be warmed and swabbed frequently if any hazard is involved. Caps, gowns and face masks are required under good manufacturing practice (GMP) conditions that are necessary under normal conditions particularly when working with laminar flow. Long hair should be tied at the back when working aseptically on an open bench. It is best to avoid talking and when the person doing the tissue culture has some infection, it is better to wear a face mask or still better to avoid performing the tissue culture.

ASEPTIC HANDLING TECHNIQUES

Aseptic technique encompasses all aspects of environmental control, personal hygiene, equipment and media sterlization and associated quality control procedures needed to ensure that a procedure is, indeed performed with aseptic technique. Most cell culture work is carried out using a horizontal laminar-flow clean bench or a vertical laminar-flow biosafety cabinet. When applicable, use of presterilized, disposable labware is recommended.

Swabbing

Swab down the work surface with 70% alcohol before and during work. Swab bottles as well especially those coming from cold storage or water bath or incubator. In case of any spill, spread a solution of 70% alcohol and swab immediately with non-linting wipes.

Capping

Deep screw caps prefer to stop us, although care must be taken when washing caps to ensure that all detergent is rinsed from rubber liners. The screw cap can be covered by the aluminium foil to protect the neck of the bottle from sedimentary dust.

Plasticware and consumables

Almost every type of cell culture vessel, together with support consumables such as tubes and pipettes, are commercially available as single-use, sterile-packed plasticware. The use of such plasticware is more cost-effective than recycling glassware, and enables a higher level of quality assurance and avoids the need for validation of cleaning and sterilization procedures. Plastic tissue culture flasks are usually treated to provide a hydrophilic surface to facilitate attachment of anchorage-dependent cells.

Centrifuges and Incubators

As discussed in Chapter 2.

CO$_2$ Incubator

The cells are grown in an atmosphere of 5–10% CO_2 because the medium used is buffered with sodium bicarbonate/carbonic acid and the pH must be strictly maintained. Culture flasks should have loosened caps to allow for sufficient gas exchange. Cells should be left out of the incubator for as little time as possible and the incubator doors should not be opened for very long. The humidity must also be maintained for those cells growing in tissue culture dishes, so a pan filled with water is kept filled all the time.

Microscopes

Inverted phase-contrast microscopes are used for visualizing the cells. Microscopes should be kept covered and the lights turned down when not in use. Before using the microscope or whenever an objective is changed, check that the phase rings are aligned.

When first developing your aseptic technique you must always be thinking of sterility. Eventually it will become second nature to you. Mastering good aseptic technique will save you considerable frustration in the labs to follow. Furthermore, the same principles for good aseptic technique also minimize biohazard risk to the investigator when infectious organisms or dangerous chemicals are used.

"EYEBALLING" THE CULTURES

Before doing anything with a culture, its general "health" and appearance should be evaluated. This can be done quickly and quantitatively by making the following observations:

1. Check the pH of the culture medium by looking at the colour of the indicator, phenol red. As a culture becomes more acid, the indicator shifts from red to yellow-red to yellow. As the culture becomes more alkaline, the colour shifts from red to fuchsia (red with a purple tinge). As a generalization, cells can tolerate slight acidity better than shifts in pH above pH 7.6.

2. Check the cell attachment. Are most of the cells well-attached and spread out? Are the floating cells dividing cells or dying cells which may have an irregular appearance?

3. The growth of a culture can be estimated by following it toward the development of a full cell sheet (confluent culture). By comparing the amount of space covered by cells with the unoccupied spaces, you can estimate per cent confluence.

4. Cell shape is an important guide. Round cells in an uncrowded culture is not a good sign unless these happen to be dividing cells. Look for doublets or dividing cells. Get to know the effect of crowding on cell shape.

5. Look for giant cells. The number of giant cells will increase as a culture ages or declines in "well-being." The frequency of giant cells should be relatively low and constant under uniform culture conditions.

6. One of the most valuable guides in assessing the success of a "culture split" is the rate at which the cells in the newly established cultures attach and spread out. Attachment within an hour or two suggests that the cells have not been traumatized and that the *in vitro* environment is not grossly abnormal. Longer attachment times are suggestive of problems. Nevertheless, good cultures may result even if attachment does not occur for four hours.

7. Keep in mind that some cells will show oriented growth patterns under some circumstances while many transformed cells, because of a lack of contact inhibition, may "pile up" especially when the culture becomes crowded. Get to recognize the range of cell shapes and growth patterns exhibited by each cell line.

COMMON TERMINOLOGY IN ASEPTIC TECHNIQUE

Antimicrobial   An agent or action that kills or inhibits the growth of microorganisms.

Antiseptic   A chemical agent that is applied topically to inhibit the growth of microorganisms.

Asepsis Prevention of microbial contamination of living tissues or sterile materials by excluding, removing or killing microorganisms.

Autoclave A steam sterilizer consisting of a metal chamber constructed to withstand the pressure that is required to raise the temperature of steam to the level required for sterilization. Early models were termed "autoclaves" because they were fitted with a self-closing door.

Bactericide A chemical or physical agent that kills vegetative (non-spore-forming) bacteria.

Bacteriostat An agent that prevents multiplication of bacteria.

Commensals Non-pathogenic microorganisms that are living and reproducing as human or animal parasites.

Contamination Introduction of microorganisms to sterile articles, materials or tissues.

Disinfectant An agent that is intended to kill or remove pathogenic microorganisms, with the exception of bacterial spores.

Pasteurization A process that kills non-spore-forming microorganisms by hot water or steam at 65–100°C.

Pathogenic A species of microorganism that is capable of causing disease in a susceptible host.

Sanitization A process that reduces microbial contamination to a low level by the use of cleaning solutions, hot water or chemical disinfectants.

Sterilant An agent that kills all types of microorganisms.

Sterile Free from microorganisms.

Sterilization The complete destruction of microorganisms.

DO'S AND DON'TS OF CELL CULTURE

Given below are a few of the essential "do's and don'ts" of cell culture. Some of these are mandatory, e.g., use of personal protective equipment (PPE). Many of them are common sense and apply to all laboratory areas. However some of them are specific to tissue culture.

THE DO'S

1. Use personal protective equipment. (laboratory coatgown, gloves and eye protection) at all times. In addition, thermally insulated gloves, full-face visor and splash-proof apron should be worn when handling liquid nitrogen.

2. Always use disposable caps to cover hair.

3. Wear dedicated PPE for tissue culture facility and keep separate from PPE worn in the general laboratory environment. The use of different coloured gowns or laboratory coats makes this easier to enforce.

4. Keep all work surfaces free of clutter.

5. Correctly label reagents including flasks, medium and ampoules with contents and date of preparation.

6. Only handle one cell line at a time. This common-sense point will reduce the possibility of cross contamination by mislabelling, etc. It will also reduce the spread of bacteria and mycoplasma by the generation of aerosols across numerous opened media bottles and flasks in the cabinet.

7. Clean the work surfaces with a suitable disinfectant (e.g., 70% ethanol) between operations and allow a minimum of 15 minutes between handling different cell lines.

8. Wherever possible maintain separate bottles of media for each cell line in cultivation.

9. Examine cultures and media daily for evidence of gross bacterial or fungal contamination. This includes medium that has been purchased commercially.

10. Quality control all media and reagents prior to use.

11. Keep cardboard packaging into a minimum in all cell culture areas.

12. Ensure the incubators, cabinet, centrifuges and microscopes are cleaned and serviced at regular intervals.

13. Test cells for mycoplasma on a regular basis.

THE DON'TS

1. Do not continuously use antibiotics in culture medium as this will inevitably lead to the apperance of antibiotic resistant strains and may render a cell line useless for commercial purposes.

2. Don't allow waste to accumulate particularly within the microbiological safety cabinet or in the incubators.

3. Don't have too many people in the lab at any one time.

4. Don't handle cells from unauthenticated sources in the main cell culture suite. They should be handled in quarantine until quality control checks are complete.

5. Avoid keeping cell lines continually in culture without returning to frozen stock.

6. Avoid cell culture becoming fully confluent. Always sub-culture at 70–80% confluency or as advised on ECACC's cell culture data sheet.

7. Do not allow media to go out of date. Shelf life is only 6 weeks at +4°C once glutamine and serum is added.

8. Avoid water baths from becoming dirty.

REVIEW QUESTIONS

1. What are the parameters that are looked into for maintaining aspetic environment in a cell culture laboratory?

2. Enumerate different aspetic handling techniques in a cell culture laboratory.

5

CULTURE ENVIRONMENT AND CULTURE MEDIA

In cell culture, it is required to provide an environment that mimics, to the greatest extent possible, the *in vivo* environment of that specific cell type. The cell culture incubator, the culture dish or apparatus, and the medium, together create this environment *in vitro*. They provide an appropriate temperature, pH, oxygen, and CO_2 supply, surface for cell attachment, nutrient and vitamin supply, protection from toxic agents, and the hormones and growth factors that control the cell's state of growth and differentiation. Clearly, this is not a simple system.

PHYSICOCHEMICAL FACTORS OF CELL CULTURE

pH

Most cell lines grow well at pH 7.4. Although optimum pH for different cell strains varies, fibroblast cell lines grow well at pH 7.4 to 7.7 and transformed cells at 7.0 to 7.4. Phenol red is the commonly used pH indicator in the media. The colour of phenol red at different pH is as follows.

Colour	pH
Lemon yellow	Below 6.5
Orange	7.0
More pink	7.6
Yellow	6.5
Red	7.4
Purple	7.8

Carbon Dioxide and Bicarbonates

CO_2 in the gas phase dissolves in the medium, establishes equilibrium with the HCO_3^- ions and thus lowers the pH. Thus pH of the culture is maintained by a bicarbonate–CO_2 buffer system.

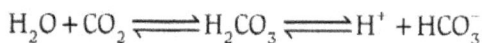

$$H_2O + CO_2 \rightleftharpoons H_2CO_3 \rightleftharpoons H^+ + HCO_3^-$$

The pK_a of the buffering system is 6.3, which is adequate but not ideal for maintaining cultures at pH 7.4, the optimum for the usual cell growth. The buffer equilibrium in the liquid phase is dependent on the presence of CO_2 in the gas phase. For pH control, an enriched CO_2 atmosphere is provided in the incubator. Normally a concentration of 24 mM bicarbonate maintains equilibrium with CO_2 at a partial pressure of the gas phase of 40 mm Hg (which corresponds to 5% CO_2).

Buffer

In addition to the weak bicarbonate buffer system, in some cases, other buffering systems will be incorporated. The commonly used one is a strong buffer HEPES (*N*-2-hydroxyethyl piperazine-*N*´-2-ethane sulphonic acid) in the pH range of 7.2 to 7.6. It is used at 10 to 20 mM concentration.

Oxygen

Cultures vary in their oxygen requirement, especially between organ and cell culture. Although atmospheric or lower oxygen tension is preferable for most cell cultures, some organ cultures require up to 95% O_2 in the gas phase. This high requirement is due to low diffusibility of oxygen. Oxygen diffusion may also be a limiting factor in porous microcarriers. Since most dispersed cell cultures rely on glycolysis for energy production, it will do better in low oxygen tension.

Osmolality

Most cultured cells have a fairly wide tolerance for osmotic pressure. The osmolality of human plasma is about 290 mosmol/kg. Osmolality is usually measured by depression of the freezing point or elevation of the vapour pressure of the medium. The addition of HEPES and drugs dissolved in strong acids and subsequent neutralization can all markedly affect osmolality.

Temperature

The optimal temperature for cell culture is dependent on 1) the body temperature of the animal from which the cells are obtained. 2) any anatomic variation in temperature (temp. of skin and testis will be lower than the rest of the body parts). Thus the temperature recommended for most human and warm-blooded animal cell lines is 37°C, close to body temperature.

Viscosity

The viscosity of a culture medium is influenced mainly by the serum content and in most cases will have little effect on cell growth. Viscosity becomes important whenever a cell suspension is agitated or when cells are dissociated after trypsinization.

Surface Tension and Foaming

The effect of foaming is important in large-scale culture vessels. Foaming will enhance the chance of contamination and also limit gaseous diffusion. The addition of a silicone antifoam or pluronic F68, 0.01–0.1%, helps prevent foaming in this situation by reducing surface tension and may also protect cells against shear stress from bubbles.

CELL CULTURE MEDIA

In the initial attempts, cell culture was performed in natural media based on tissue extract and body fluids, such as chick embryo extract, serum, lymph, etc. With the propagation of cell lines, the demand for larger amounts of medium of more consistent quality lead to the introduction of chemically defined media based on analysis of body fluids and nutritional biochemistry. **Eagle's basal medium** and subsequently **Eagle's Minimal Essential Medium** (MEM) became widely adapted, variously supplemented with calf, human, or horse serum, protein hydrolysates and embryo extracts.

Isolation and propagation of cells of a specific lineage may require a selective serum-free medium, whereas cells grown for the formation of products and hosts for viral propagation for non-cell-specific molecular study rely mainly on Eagle's MEM, Dulbecco's modification of Eagles medium (DMEM) or increasingly RPMI 1640 medium, supplemented with serum. Many industrial-scale production techniques use serum-free media to facilitate downstream processing and reduce the risk of adventitious infectious agents.

The nutrient mixture is the cornerstone of cell culture. Having the correct nutrient mixture can often be the determining factor in the failure or success of growing a cell *in vitro*. The medium provides essential nutrients that are incorporated into dividing cells, such as amino acids, fatty acids, sugars, ions, trace elements, vitamins and cofactors, and ions and molecules necessary to maintain the proper chemical environment for the cell. Some components may perform both roles; for example, sodium bicarbonate used as a carbonate source may also play the important role of maintaining the appropriate pH and osmolality. The medium contains all or part of the buffering system required to maintain a physiological pH and should provide the appropriate osmolality of the cells.

Some commonly used culture media are listed in Table 5.1.

Table 5.1   Commonly used culture media

Media	Composition/Purpose
BME (Eagle's basal medium)	Originally designed for mouse L and HeLa cells.
EMEM (Eagle's minimal essential medium)	Used for a wide variety of cell lines.
DMEM (Dulbecco's modification of Eagle's medium)	Medium has 4X amino acid and the vitamin conc. of BME.
GMEM (Glasgow's modification of Eagle's medium)	Medium has 2X amino acid and vitamin conc. of BME.
RPMI 1640 (Roswell Park Memorial Institute medium)	Used for lymphocyte and hybridoma cultures.
Leibovitz	Used for fibroblast growth in the absence of a CO_2-atmosphere.
Ham's F-12	Has a complex composition and is used for a variety of cell lines.
199	An extremely complex medium (61 components) which can support cell growth without serum.

COMPONENTS OF A TYPICAL CULTURE MEDIUM

Carbohydrates   Glucose is used in most formulations to act as a source as well as precursor for biosynthesis of ribose needed for nucleic acid synthesis. Fructose can also be used as an alternative which results in decreased lactic acid production and thus more stable culture pH.

Amino acids   There are two types of amino acids—essential amino acids which are those not manufactured by the cell plus cysteine and tyrosine, and non-essential amino acids which are often added to the medium to compensate for a particular cell type which is unable to manufacture them or if they deplete rapidly. Amino acids are included at a concentration of 0.1 to 0.2 mM as a source of precursors of protein synthesis. The richer media contain both "essential" and "non-essential" amino acids.

Salts   Salts are included so that the solution is isotonic and has no imbalance with the intracellular contents. The osmolarity of the standard culture medium is approximately 300 mOsm/l and is optimal for most of the cell lines. Cells can normally tolerate variations within 10% of this value. However care should be taken while adding supplements because they may adversely affect osmolarity. Amino acids and glucose, as well as ions such as NaCl, contribute to the osmolality of the medium. **Bicarbonate** is usually used as a buffer system in conjunction with a gaseous atmosphere of 5–10% CO_2 provided by the incubator. This allows cultures to be maintained at a pH of 6.9 to 7.4. The disadvantage of bicarbonate–CO_2 buffer system is that cultures may became alkaline very quickly when removed from the incubator. To prevent this or to increase the buffering capacity of the culture, the organic buffer **HEPES** (pK_a = 7.3 at 37°C) may be added at a concentration of 10–20 mM. In the presence of HEPES, the CO_2 level can be reduced to around 2% with a concomitant decrease in bicarbonate concentration.

Vitamins and hormones   These are present at relatively low concentrations (μ-molar) and are utilized as metabolic cofactors. Many media contain the common vitamins such as niacin, folic acid, riboflavin, inositol, thiamine, and so forth. While these vitamins are essential to continued cell replication, a detrimental effect may not be seen until several cell doublings are produced after their removal from the medium. Other vitamins such as vitamins D (1,25-dihydroxycholecalciferol), C (ascorbic acid), E (α-tocopherol), and A (retinol, retinoic acid) are not commonly added to media formulations because they are unstable in solution. (Even if these are in the medium formula, they may not be active by the time the medium gets to the cells.) However, these may prove beneficial or even essential for some cell types and should be added separately. They may also be involved in maintaining the differentiated state of the cell, in regulating cell functions, or acting as antioxidants.

Phenol red   Most media contain phenol red as a pH indicator. This is very helpful in rapidly assessing the pH of the medium of all the cultures in an incubator. Phenol red can be added to media if it is not part of the medium powder or if a more obvious colour is desired. It should be noted that phenol red has weak oestrogenic activity, which may be a consideration with some cells. At lower pH, the phenol red becomes orange (pH 7.0) or yellow (pH 6.5). An overnight change in colour of the culture from red to yellow usually indicates bacterial contamination.

Other components   Media also contain lipids. Most contain a mixture of fatty acids, and some contain more complex lipids (e.g., cholesterol). Some media formulations, such as medium 199, contain detergents (e.g., Tween 80) to help emulsify the lipids. These detergents can be toxic to some types of cells, particularly in serum-free medium. Some media contain macromolecules such as thymidine, adenosine, and hypoxanthine that can be synthesized by cells *in vitro*. Adding more of these to the medium may nonetheless improve the growth of some cells by maintaining an appropriate pool size of precursors in the cells. Medium may also contain antioxidants or reducing agents (or these might be added separately).

Serum supplementation   Serum is normally added to the medium at a concentration of 10% (v/v) to promote cell growth. Cow (bovine) or horse serum is most commonly used, and foetal calf serum (FCS) is being considered particularly effective because of its high content of embryonic growth factor. Serum is filtered by 0.1 μm pore size filter. Each batch is tested for a variety of microbial chemical contaminants, including bacteria, fungal spores and specific viruses, depending on the source of the serum.

Most media [e.g., **minimal essential (ME) medium, Dulbecco's Modified Eagle's (DME) medium**] were developed specifically for use with serum supplementation and high density growth of cells. In contrast, others such as **Ham's nutrient mixtures** F12, F10, and the **Molecular Cellular Developmental Biology (MCDB)** series of media were tailored specifically for growing a given cell type at low density with a minimal amount of undefined protein added, so as to study the effects of the nutrient components of the media. **RPMI** (Rosewell Park Memorial Institute) series of media is also widely used in growing cells. **Leibovitz L-15 medium** is designed to grow cells in equilibrium with air rather than CO_2 and is useful when CO_2 incubators are not available (e.g., the teaching laboratory), or when cells are shipped or handled extensively outside the incubator (for example, during a long tissue dissociation protocol).

Alternative to serum   Although widely used, serum supplementation has many disadvantages.

1. It is chemically undefined and variations between batches can result in inconsistent promotion of cell growth.
2. It is expensive. FCS accounts for 70–80% of the cost of some formulations. This is an important consideration in large-scale cultures.
3. The proteins in the serum can compromise the extraction and purification procedures for cell-secreted proteins.

For these reasons, considerable effort has been made to develop a supplement of hormones and growth factors to replace serum. The serum-free supplements are specific for particular cell type. Commonly used ingredients for these supplements include insulin, transferrin, ethanolamine and selenite. Pre-prepared serum–free media supplements are commercially available but, although these are chemically defined, the manufacturers do not usually reveal the content. For an extended study on a particular cell line, it is often worth spending some time in developing a serum-free formulation based on a modification of a standard published recipe. When developing a serum-free formulation, it is very important to start with an optimal basal medium—rich formulations such as DMEM and Ham's F-12 (50:50) have been found particularly suitable.

More recently, vendors are supplying "special-use media" to grow a stated cell under special conditions. These sometimes contain undisclosed hormones, growth factors, or undefined protein components. Therefore, these cannot be considered "defined," although they may work well for some applications. Other such media are supplied with a defined supplement mix that must be added before use.

Antibiotic These are included in media for short-term cultures in order to reduce the risk of contamination. The optimal concentration of antibiotics should be determined empirically bearing in mind that they may be cytotoxic. Antibiotics are often used in combination in culture media and the following cocktail can be recommended for general use—Penicillin-G (100 U/ml) to inhibit the growth of gram-positive bacteria, streptomycin (5 mg/l) to inhibit the growth of gram-positive and gram-negative bacteria and amphotericin B (25 mg/l) as an antifungal agent.

CLASSIFICATION OF CELL CULTURE MEDIA

Classical media

1. DMEM, low Glucose (1 g/l)
2. DMEM, high Glucose (4,5g/l)
3. Dulbecco's MEM (DMEM), Thermostable Media
4. DMEM/Ham's F-12
5. Ham's F-10
6. Ham's F-12
7. Iscove's modified DMEM (IMDM)
8. L-15 Medium (Leibovitz's)
9. Medium 199 with Earle's Salts
10. Medium 199 with Hanks' Salts
11. MEM Alpha Modification
12. Minimal Essential Medium Eagle (MEM)
13. RPMI 1640
14. RPMI 1640, Thermostable Media
15. DMEM Ready Mix
16. RPMI 1640 Ready Mix
17. Quantum 3–21
18. Quantum PBL
19. Ham's F-10 Ready Mix

Neuronal media

20.   Neuronal Base Medium

21.   Neuronal Base Medium AD

Stem cell media

22.   CollagenStem Kit

23.   MethoStem Medium

24.   MesenchymStem Medium

Cell type specific media

25.   Endothelial Cell Medium

26.   Macrophage Medium

27.   Hybridoma Express

28.   InsectExpress Sf9-S2

29.   TC 100 Insect Medium

30.   Quantum 101 Complete Medium for HeLa Cells

31.   Quantum 263 Complete Medium for Tumour Cells

32.   Quantum 286 Complete Medium for Epithelial Cells

33.   Quantum 333 Medium for Fibroblasts

SELECTION OF APPROPRIATE MEDIUM

If a new cell line is brought into the laboratory, determine what medium is recommended for its growth. This information can be obtained from the same source as the cells. If the recommended medium is incompatible with the CO_2 settings on the incubator used for other cells grown in the laboratory, or is not commonly prepared in the laboratory, one might wish to change the growth medium. It is best to initially grow the cells in their original medium and compare this with the more convenient medium after a passage or two in each. If the growth rate and morphology of the cells look the same, then a medium switch can be made. However, keep in mind when trying to repeat published data that cells grown in a different medium may respond differently in some parameters measured. For example, oestrogen-requiring cells previously grown in F-12 medium and then switched to DME medium may have a diminished response to exogenously added oestrogens in this medium because of the weak oestrogenic activity of the phenol red, which is present at a much higher concentration in the DME medium.

If one wishes to grow a primary cell in culture and no published data exists on growing that cell type *in vitro*, or one wishes to grow the cells in a different manner (e.g., with defined supplements rather than serum), it is best to screen several of the commercially available media before deciding on the one that is best for that particular use. This can be done by obtaining five to ten candidate media powders from a supplier, preparing them all in the laboratory as described below using the same water and supplementary components, and doing a direct comparison of cell growth in the different conditions. If end points other than cell growth are important, measure these too in each of the media. Carry the cells in the selected medium for several passages and freeze them in this medium for future use.

MEDIA PREPARATION

Media can be purchased as prepared liquid media, made up in the laboratory from dried powders containing most of the components of the nutrient mixtures, or prepared in the laboratory from individual stocks of the individual components or groups of the components. Purchasing liquid medium is not recommended, especially for serum-free culture work. Medium components deteriorate with time, and do so faster in solution. Some necessary components break down and are lost, others create toxic breakdown products or oxidize to toxic components. While it varies from cell to cell and with serum-supplemented or serum-free media, it is found that 2 weeks is a safe storage time for serum-free media, or longer if serum is added when the medium is prepared. Outdated medium can be used for washing cells or preparing tissues for primary culture. Some prepared liquid media can be frozen. Those that form a precipitate when thawed should not be frozen. In any case, it is always safe to store the prepared powdered medium and make liquid medium in the laboratory on a regular basis.

Powdered nutrient mixtures generally have a shelf life of a year or more if stored in moisture-proof, air-tight containers in the dark. If the laboratory does not use large volumes of media, the 1-litre packages are convenient. Preparing medium in the laboratory from components immediately before use is obviously the best way to insure that the medium contains the desired components in the desired form. However, most laboratories will find that preparation of medium from commercial powdered nutrient mixtures and a limited storage of the prepared media in a light-tight refrigerator will be adequate for their needs. This is also less costly than purchasing prepared media, especially when the cost of filters and so forth can be spread over large-volume use.

TESTING MEDIA AND COMPONENTS—QUALITY CONTROL

It is best to prepare media from commercially available powdered nutrient mixtures as outlined above. It is important to keep good records and do quality control testing of reagents used in making the medium. Generally one set of glassware is kept exclusively for preparing media. This is rinsed well with distilled water but not washed with detergent between each use. This avoids the possibility of any detergent residue getting into the medium. The sodium bicarbonate and other reagents that are used in media should be used only for media. When weighing out these reagents, a disposable tongue depressor and weigh boat should be used. This avoids contaminating these reagents with other, potentially toxic chemicals that may be in use in the laboratory.

The major component of the medium is water. Water purity is very important for good-quality medium. It is a good practice to do a plating efficiency experiment, since cells at low density are more sensitive to toxic components, while high-density growth would be better able to detect a medium deficient in nutrients.

TROUBLESHOOTING MEDIUM PROBLEMS

Even when these precautions are followed, there will come a time when problems arise that require troubleshooting. This is when meticulous testing and record keeping pay off. Follow the steps below to identify and eliminate the problem:

1. Talk to all persons using the culture facility. Determine whether the problem is being experienced in many different cell lines or only a few; by all users or only a few.

2. When did the problem start? Determine the earliest date that anyone thought they might have a problem.

REVIEW QUESTIONS

1. Enumertate different physicochemical factors affecting cell culture.

2. What are the different components of cell culture media?

3. How are the cell culture media now classified?

6

MAINTENANCE AND STORAGE OF CULTURE

There are two common methods of cell preservation at low temperature: **hypothermic preservation** at temperatures above freezing and **cryogenic preservation** at temperatures below freezing. Hypothermic preservation allows short-term storage of tissues and organs prior to further investigation. Cryogenic preservation is used for the establishment of cell banks to ensure reproducible results in research by stabilizing living cells. Using cryopreservation media, the cells are protected from damage which can occur during freezing and warming processes.

HYPOTHERMIC PRESERVATION

Hypothermic preservation at 2 to 8°C is suitable for short-term preservation of cell culture. Hypothermic preservation is a procedure to stabilize tissues and organs at +2°C to +8°C. The goal of cold storage is to induce hypothermia in order to suppress the rate of cell deterioration. Usually hypothermic preservation is done in the presence of specific media components which prevent cell decay by reduction of free oxygen radicals and inhibition of different proteases and nuclease. Several commercial hypothermic preservation media are available. They are based on saccharides and derivates, salts, energy sources, reducing agents and other stabilizers. Also, an optimal preservation solution contains components to reduce cell-swelling (colloids), buffer to maintain pH balance, antioxidants to scavenge free oxygen radicals and ATP precursors to provide energy upon reperfusion.

APPLICATIONS

The cold storage of tissues and organs with optimized hypothermic preservation media is a prerequisite for successful organ transplantation. For successful tissue and organ transplantation, the individual cells must be preserved with the highest viability to maintain the functionality of the whole organ. In contrast to single growing cells, only the outermost cells organized in tissues and organs are provided with essential nutrient components, whereas the innermost cells are supplied with nutrients only by diffusion. In order to reduce the damaging effects of proteases, nucleases and free radicals, the solution must contain specific inhibitors and components for basic maintenance of the cellular metabolism. Only these specific components allow a high diffusion rate into innermost cell layers assuring optimal supply and preservation.

CRYOGENIC PRESERVATION

Cryopreservation is a process used to assure an adequate stock of cells and serve as a back-up in case of any contamination of the working cell lines. With a stock of frozen cells, constant quality can be guaranteed without change of cellular and genetic features. Only under cryopreserved conditions the cells can maintain homogeneous populations because the cellular metabolism is reduced to a minimum without any genetic drifts.

Cryopreservation is a procedure to stabilize cells at cryogenic temperatures for long-term storage in liquid nitrogen. During the freezing process the usage of cryoprotective additives protects fragile membranes and organelles of the cells by minimizing the damage associated with ice-crystal formation. Classical cryopreservation media contain dimethyl sulphoxide (DMSO) with stabilizers such as FBS or methylcellulose which increase the viability after thawing.

The success of the freezing process depends on four critical areas:

1. Proper handling and gentle harvesting of the cultures
2. Correct use of the cryoprotective agent
3. A controlled rate of freezing
4. Storage under proper cryogenic conditions

PRINCIPLE

The basic principle of successful cryopreservation is a slow freeze and quick thaw. The freezing process involves complex phenomena. Since water is the major component of all living cells and must be available for the chemical processes of life to occur, cellular metabolism stops when all water in the system is converted to ice. Ice forms at different rates during the cooling process. During slow cooling, freezing occurs external to the cell before intracellular ice begins to form. As ice forms, water is removed from the extracellular environment and an osmotic imbalance occurs across the cell membrane leading to water migration out of the cell. The increase in solute concentration both outside the cell, and intracellularly, can be detrimental to cell survival. If too much water remains inside the cell, damage due to ice-crystal formation and recrystallization during warming can occur. This is a suitable method for long-term storage of myeloma cells, hybridoma cells, T cells, and other mammalian cell lines.

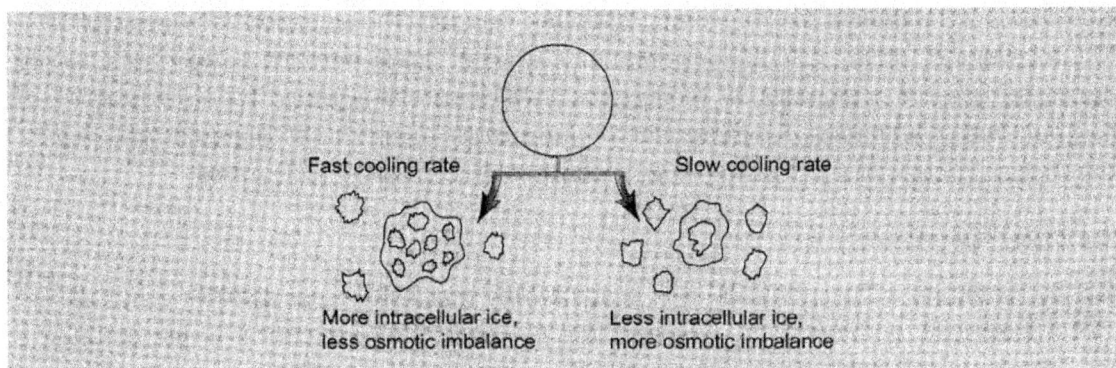

Figure 6.1 Preservation and cell freezing (Hay *et al.*, 2000)

FROZEN CELL STORAGE AND LIQUID NITROGEN REFRIGERATORS

Frozen cells should be stored at –180°C. This is liquid nitrogen temperature. Cell storage tanks are specially designed to handle these low temperatures. A good tank should have a long "holding time" (the time taken for all the liquid nitrogen to evaporate if the tank is not refilled), hold a reasonably large number of vials, and have a good inventory system for storing and finding the vials. Larger tanks use less liquid nitrogen in proportion to their storage capacity than smaller tanks. Optimally, cells should be stored in the vapour phase rather than the liquid phase, but this increases the risk of the tank going dry.

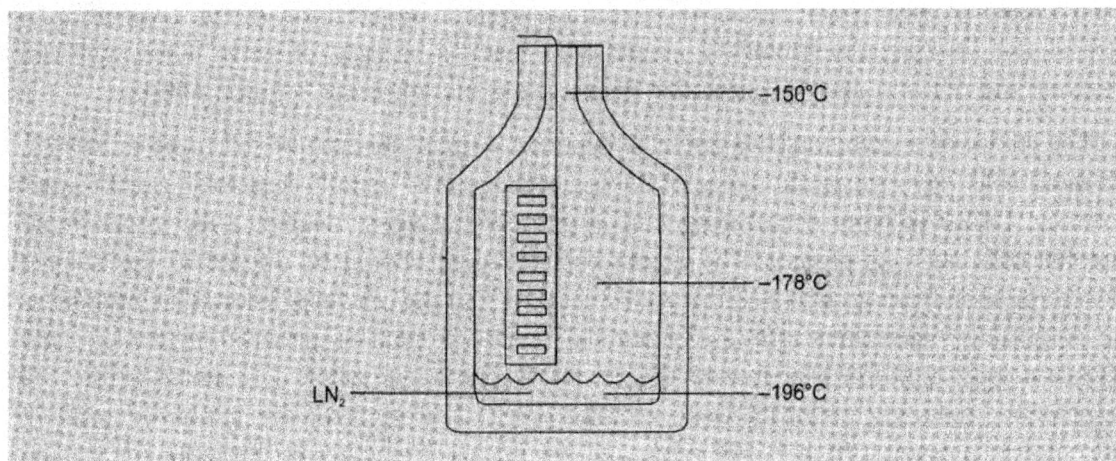

Figure 6.2 Liquid nitrogen container, showing different temperatures and the position of the canister (Hay *et al.*, 2000)

SLOW FREEZING APPARATUS

The optimum freezing rate for cell lines (for most cells about –1°C/min) can be achieved through use of apparatus varying in complexity from a tailor-made Styrofoam box to a completely programmable freezing unit. A cooling rate of –1°C reduction is warranted in a –70°C mechanical freezer. Alternatively, manufacturers of liquid nitrogen refrigerators supply adapted refrigerator neck plugs, at modest cost, which can be adjusted for slow freezing of small numbers of ampoules. For those who produce larger quantities of ampoules and require more precise control of the freezing rate, a controlled rate freezer is required.

RECORD KEEPING

Since by its nature a frozen vial of cells is something that will be put away and not looked at for a long time, perhaps years, labelling and record-keeping are critical to maintaining a good laboratory cell bank. Each vial of cells should be labelled with the complete name of the cells, the passage number, the date, and the name or initials of the person growing and freezing the cells (Figure 6.3). A laboratory-wide system should be established to maintain records of what cell lines are frozen, how many vials are frozen, who froze them, why they were frozen at that time, and where they are stored.

CRYOPROTECTIVE AGENTS

Many compounds have been tested as cryoprotective agents, either alone or in combination, including sugars, serum and solvents. Although there are no absolute rules in cryopreservation, glycerol and DMSO have been widely used and seem to be most effective. Cryoprotective agents serve several functions during the freezing process. Freezing point depression is observed when DMSO is used which serves to encourage greater dehydration of the cells prior to intracellular freezing. Cryoprotective agents also seem to be most effective when they can penetrate the cell and delay intracellular freezing and minimize the solution effects.

The choice of a cryoprotective agent is dependent upon the type of cell to be preserved. For most cells, glycerol is the agent of choice because it is usually less toxic than DMSO. However, DMSO is more penetrating and is usually the agent of choice for larger, more complex cells such as protists. The cryoprotective agent should be diluted to the desired concentration in fresh growth medium prior to adding it to the cell suspension. This minimizes the potentially deleterious effects of chemical reactions such as generation of heat, and assures a more uniform exposure to the cryoprotective agent when it is added to the cell suspension, reducing potential toxic effects. DMSO and glycerol are generally used in concentrations ranging from 5–10% (v/v), and are not used together in the same suspension with the exception of plant cells. The optimum concentration varies with the cell type and the highest concentration the cells can tolerate should be used. In some cases, it may be

FREEZE DATA FORM

NAME _____   DATE _____

Growth Temp.	Cryopreservation Solution	Culture		Location	Inventory		
		Age	Medium		Storage Temp.		Seed
					Gas		Working lot

MICROSCOPIC EXAM _____

PREPARATION FOR FREEZE:

    Equilibration time _____   Equilibration temp. _____

DISPENSING:

    Vial type _____   Volume per vial _____

FREEZING:

    Program rate _____

SURVIVAL:

    Prefreeze count _____ cells/ml   Total vol. frozen _____ ml   % viable _____

    Postfreeze count _____ cells/ml   Total vol. resuspended _____ ml

	Survival			NOTES
	% Rec	Purity (free from contamination)	No. pass	

COMMENTS

Figure 6.3 A freeze data form

advantageous to examine the sensitivity of the cells to increasing concentrations of the cryoprotective agent to determine the optimum.

The most commonly used cryoprotective agents serve several functions during the freezing process, e.g., freezing point depression, protection of solution effects and extra- and intracellular ice-crystal formation, but are often toxic at high concentrations.

ADVANTAGES OF CRYOPRESERVATION

The aim of cryopreservation is to enable stocks of cells to be stored to prevent the need to have all cell lines in culture at all times. It is invaluable when dealing with cells of limited lifespan. The other main advantages of cryopreservation are:

- Reduced risk of microbial contamination
- Reduced risk of cross contamination with other cell lines
- Reduced risk of genetic drift and morphological changes
- Work conducted using cells at a consistent passage number (refer to cell banking section below)
- Reduced costs (consumables and staff time)

GENERAL PROTOCOL FOR FREEZING

1. Harvest cells from late log phase or early stationary growth phase. Scrape cells from the growth surface if they are anchorage-dependent. Centrifuge broth or anchorage-independent cultures to obtain a cell pellet, if desired (Figure 6.4).
2. Prepare pre-sterilized DMSO or glycerol in the concentration desired in fresh growth medium. When mixing with a suspension of cells, prepare the cryoprotective agents in twice (2×) the desired final concentration.
3. Add the cryoprotectant solution to the cell pellet or mix the solution with the cell suspension. Begin timing the equilibration period.
4. Gently dispense the cell suspension into vials.
5. Begin cooling the cells after the appropriate equilibration time.
    i. *Uncontrolled cooling*   Place the vials on the bottom of a –60°C freezer for 90 minutes.
    ii. *Semi-controlled cooling*   Use specific freezing container to freeze the vials in a –70°C freezer.
    iii. *Controlled cooling*   Use a programmable cooling unit to cool the cells at –1°C per minute to –40°C.
6. Remove the cells from the cooling unit and place them at the appropriate storage temperature.
7. To reconstitute, remove a vial from storage and place into a water bath at 37°C. When completely thawed, gently transfer the entire contents to fresh growth medium.

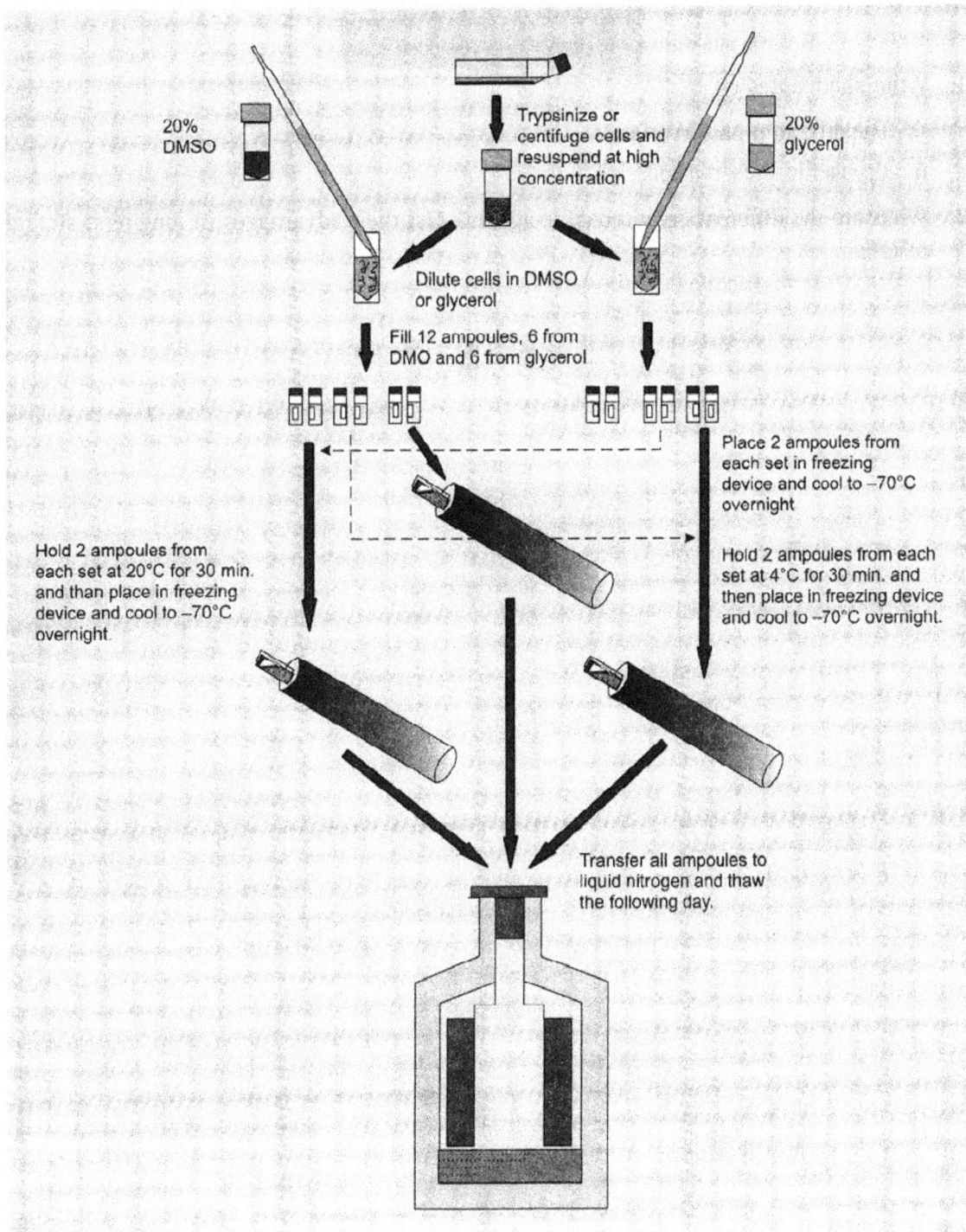

20%
DMSO

Trypsinize or
centifuge cells and
resuspend at high
concentration

20%
glycerol

Dilute cells in DMSO
or glycerol

Fill 12 ampoules, 6 from
DMO and 6 from glycerol

Place 2 ampoules from
each set in freezing
device and cool to –70°C
overnight

Hold 2 ampoules from
each set at 20°C for 30 min.
and than place in freezing
device and cool to –70°C
overnight.

Hold 2 ampoules from each
set at 4°C for 30 min. and
then place in freezing device
and cool to –70°C overnight.

Transfer all ampoules to
liquid nitrogen and thaw
the following day.

Figure 6.4 Experimental protocol for cell freezing with DMSO and glycerol (Hay *et al.*, 2000).

REVIEW QUESTIONS

1. Write short notes on

   i. Hypothermic preservation

   ii. Cryogenic preservation

2. What are the different cryoprotective agents? List their advantages in long-term storage of cells.

7

CELL LINE BANKING

CELL BANKING

It is not good practice to maintain a cell line in continuous or extended culture for the following reasons.

1. Risk of microbial contamination
2. Loss of characteristics of interest (i.e., surface antigen or monoclonal antibody expression)
3. Genetic drift particularly in cells known to have an unstable karyotype
4. Loss of cell line due to exceeding finite lifespan, e.g., human diploid cells such as MRC-5
5. Risk of cross contamination with other cell lines
6. Increased consumables and staff costs

Cryopreservation of cells plays an important role in establishing cell stocks. Cryopreserved cells assure an adequate supply in case of microbial contamination of currently used cells. By the implementation of a master cell bank (MCB) and working cell bank (WCB), all potential risk factors may be minimized. Establishing an MCB and a WCB are the first steps in the manufacturing process for biopharmaceuticals and can be prepared under serum-free conditions with defined and animal-derived component-free cryopreservation media.

All the potential risk factors may be minimized by the implementation of cell banking system. This type of system is known as a **tiered banking system** or **master cell banking system**. On initial arrival into the laboratory, a new cell culture should be regarded as a potential source of contaminants like bacteria, fungi and mycoplasma and should be handled under quarantine conditions until proven negative for such microbial contaminants.

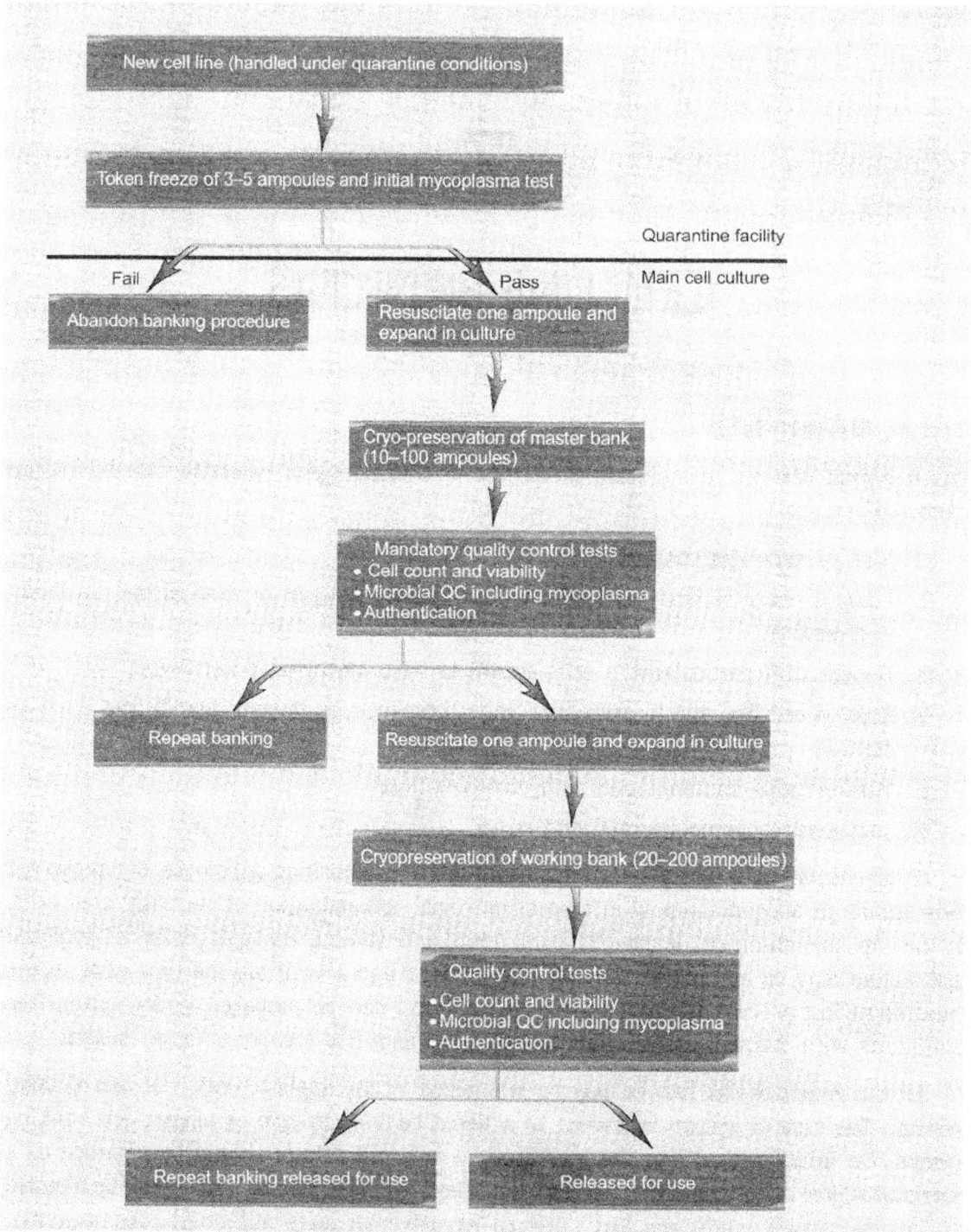

Figure 7.1   Establishment of cell line bank (Hay *et al.*, 2000)

Following initial expansion 3–5 ampoules should be frozen as a token stock before a master bank is prepared. One of the token stock ampoules should then be thawed and expanded to produce a master bank of 10–20 ampoules depending upon the anticipated level of use.

Ampoules of this bank (2–3) should be allocated for quality control comprising confirmation that the cell count and viability of the bank is acceptable and that the bank is free of bacteria/fungi and mycoplasma. Additional tests (such as viral screening and authenticity testing) may also be required. Once these tests have been completed satisfactorily, an ampoule from the Master Bank should be thawed and cultured to produce a **Working Bank**. The size of this bank will again depend on the envisaged level of demand. Quality control tests (cell count and viability and the absence of microbial contaminants) are again required prior to using the cultures for routine experimentation or production. It is also important at this stage to confirm by DNA profiling techniques that the Master and Working Banks are genetically identical.

Implementation of this banking system ensures the following.

- Material is of a consistent quality.
- Experiments are performed using cultures in the same range of passage numbers.
- Cells are in culture when required.
- The original cell line characteristics are retained.

Important cell culture collections are:

ATCC —American Type Culture Collection

HSRRB—Health Science Research Resources Bank, Japan

EACC— European Animal Cell Culture Collection

In India cell lines can be obtained from National Centre for Cell Sciences (NCCS), Pune.

BENEFITS OF CRYOPRESERVATION

Cryopreservation is invaluable when dealing with cells of limited lifespan. It allows cells to be stored at ultra-low temperature for future use without having to resort to the continuous cultivation of cell lines. Other main advantages include:

- Reduced risk of microbial contamination
- Reduced risk of cross contamination with other cell lines
- Reduced risk of genetic drift and morphological changes
- Work conducted using cells at a consistent passage number
- Reduced costs

THE REQUIREMENTS FOR SUCCESSFUL CRYROPRESERVATION

A large amount of development work has been undertaken to ensure successful cryopreservation and resuscitation of a wide variety of cell lines of different cell types. The basic principle of cryopreservation is to slow-freeze and quick-thaw. The most reliable and reproducible way to achieve a slow-freeze at a rate of –10°C to –30°C per minute is with the use of a programmable rate-controlled freezer. The cost in acquiring such equipment is often beyond the budget for the majority of research laboratories.

Cryopreservation also depends upon the use of a high concentration of serum/protein (>20% should be used and in many cases serum is used at 90%) and cryoprotectants such as dimethylsulphoxide. Both cryoprotectants help to prevent the cells from rupturing due to the formation of ice crystals. DMSO is the most common cryoprotectant used at a final concentration of 10%, however, this is not always appropriate because DMSO induces differentiation in some cell lines (e.g., HL-60). In such cases, glycerol is often used as the alternative. It is essential that immediately prior to cryopreservation, cultures should be healthy with a viability of >90% and in the log phase of growth. The latter parameter can be achieved by using pre-confluent cultures (i.e., cultures that are below their maximum cell density).

ULTRA-LOW-TEMPERATURE STORAGE OF CELL LINES

Following controlled rate freezing, cells can be cryopreserved in a suspended state for an indefinite period, provided a temperature of less than –135°C is maintained. European collection of cell culture (ECACC) strongly discourages the idea of long-term storage at –800°C. Such ultra-low temperatures can be attained only by the following. The advantages and disadvantages of each are summarized below in Table 7.1.

- Specialized electric (–135°C) freezer
- Liquid phase nitrogen
- Vapour phase nitrogen

Table 7.1 Comparison of ultra-low temperature storage methods for cell lines

Methods	Advantages	Disadvantages
Electric (−135°C) freezer	• Ease of maintenance • Steady temperature • Low running costs	• Requires liquid nitrogen back-up • Mechanically complex • Storage temperatures high relative to liquid nitrogen
Liquid phase nitrogen	• Steady ultra-low (−196°C) temperature • Simplicity and mechanical reliability	• Requires regular supply of liquid nitrogen • High running costs • Risk of cross contamination via the liquid nitrogen
Vapour phase nitrogen	• No risk of cross contamination from liquid nitrogen • Low temperatures achieved • Simplicity and reliability	• Requires regular supply of liquid nitrogen • High running costs • Temperature fluctuations between −135°C and −190°C

REVIEW QUESTIONS

1. What are the benefits of cryopreservation?

2. Write short notes on cryopreservation of cells?

8

CELL QUANTITATION METHODS

Quantitation of cell lines is essential during many instances in cell culture including the following.

1. Characterization of growth properties of different cell lines.
2. Evaluation of culture conditions.
3. Checking the consistency of primary cultures.
4. Checking the survival of cells on storage.

Growth in cell cultures is normally determined by counting cells at regular intervals.

Methods for quantitation can be classified into

1. Direct methods
2. Indirect methods

DIRECT METHODS

The two direct counting methods commonly used are microscopic method using a haemocytometer or electronic counting by a particle counter (e.g., Coulter current counter).

HAEMOCYTOMETER COUNT

The haemacytometer is the cheapest and most labour-intensive method for counting cells, but it can be used to provide data as accurate as that obtained by any other method. It provides an assessment of both total and viable cell counts. Cells can be counted before, during, and after setting up an experiment to accurately and directly quantitate and standardize an experimental condition. Moreover, the use of the dye such as trypan blue when doing haemocytometer counts gives the investigator a quantitative standard for the viability of the cells by doing a differential count of the cells that exclude trypan blue ("sort-of-sick" to viable) and those that take up the dye (irretrievably dead).

The number of cells/ml and the total number of cells are determined using the following formula:

$$Cells/ml = \frac{No.\,of\,cells\,counted}{No.\,of\,squares\,counted} \times 10^4 \times Dilution\,factor$$

The percentage of viable cells can be calculated using the following formula:

$$\%Viability = \frac{No.\,of\,viable\,cells\,counted}{Total\,no.\,of\,cells\,counted} \times 100$$

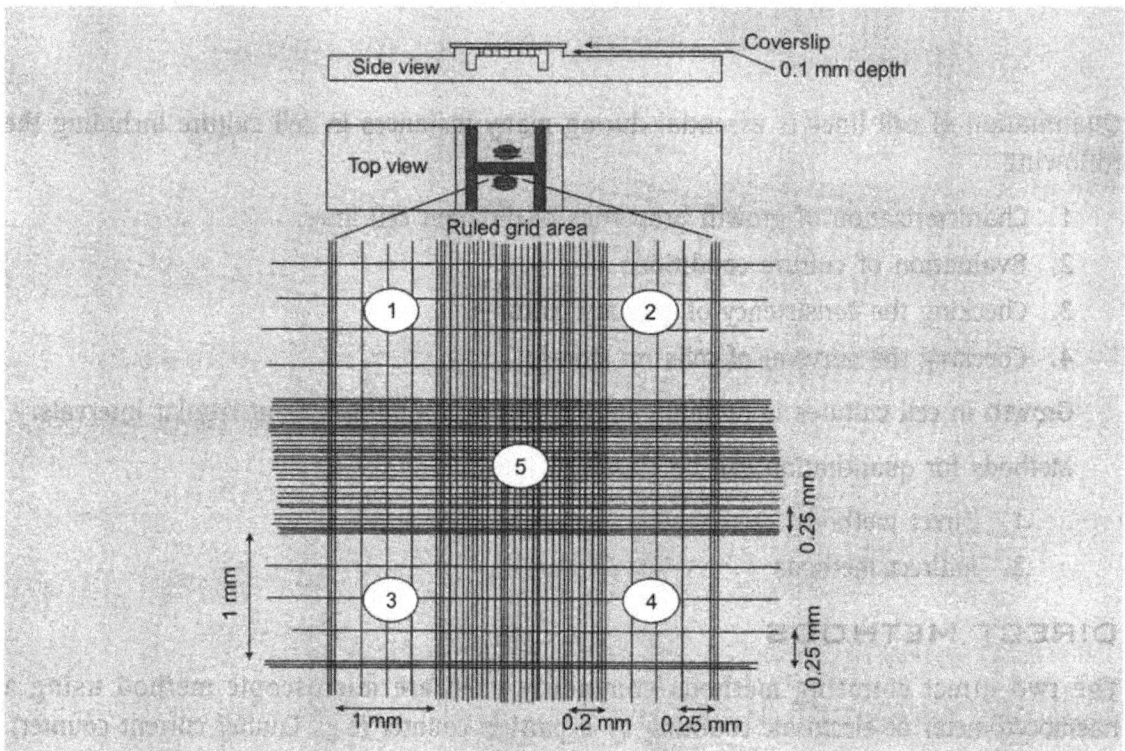

Figure 8.1   Haemocytometer, side view, top view and magnified haemocytometer grids

ELECTRONIC PARTICLE COUNTER

An **electronic cell counter** provides an easy quantitation of cells. The **Coulter counter**, manufactured by Coulter Electronics, affords a rapid, accurate, and reproducible method of total cell counting, particularly when dealing routinely with a large number of samples to count. As a cell passes through the aperture through which an electrical current is flowing, it displaces an equal volume of electrolyte. This causes a change of resistance in the path of the current and subsequently a change in the voltage. This change in voltage is directly proportional to the volume or size of the cell. Every change in voltage during the sample flow is represented as a cell count, which is then displayed on the LED read-out. The counts shown include both viable and nonviable but intact cells. The counting threshold can be set to avoid counting cell debris. Cells must be single-cell suspensions to obtain accurate cell counts with these machines. The machines will also underestimate cell number in a manner related to the cell density. As the density becomes greater, there is an increasing chance that

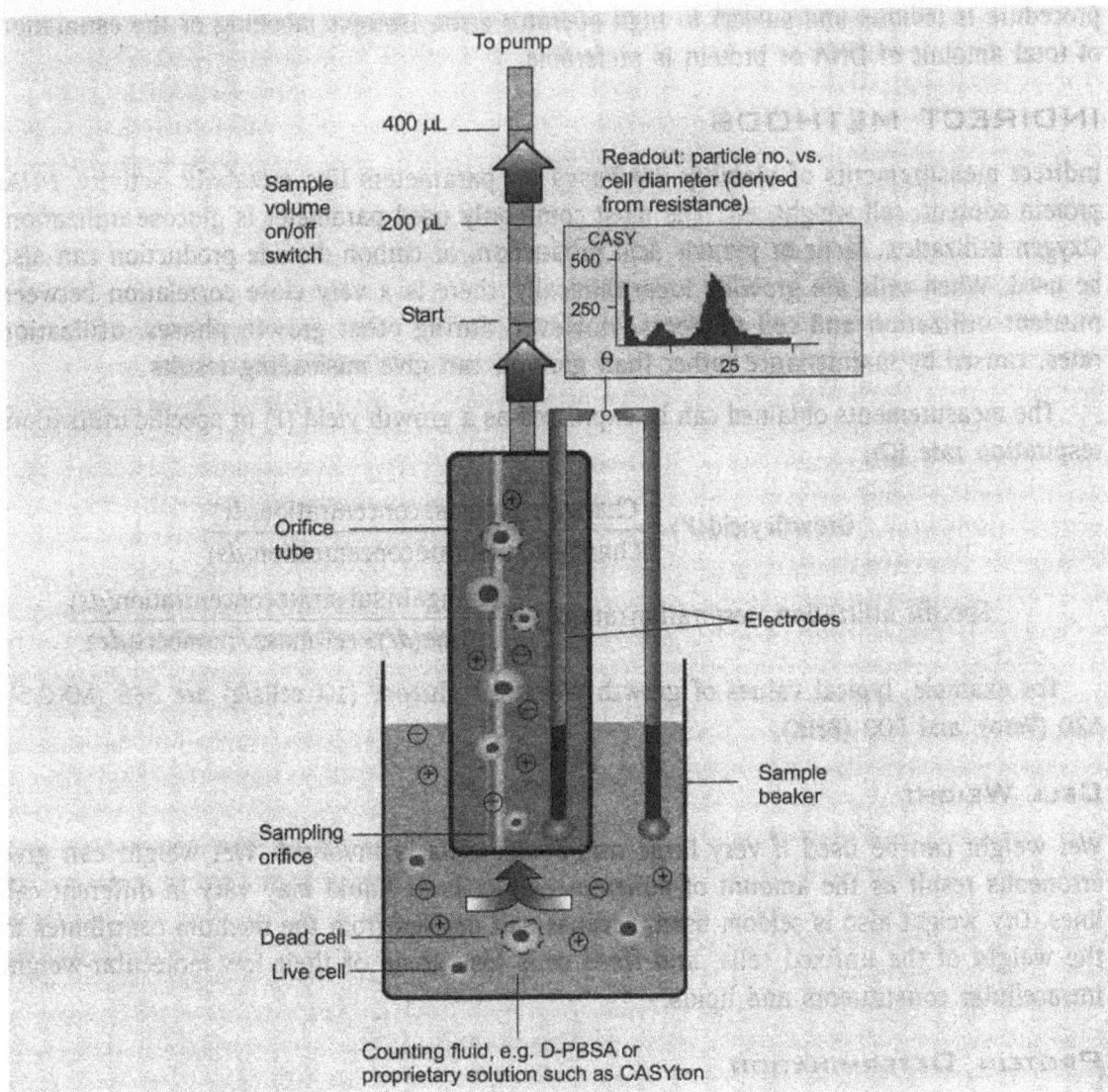

Figure 8.2 Electronic counter (Mathur and Barnes, 1998)

two cells will pass through the aperture so close together that they will only be counted as a single cell. This error can be corrected for by a mathematical formula that is used to generate the "coincidence correction tables" provided with the machines. This can be used to correct the counts manually in older model Coulter counters and is done automatically in the newer model counters.

COUNTING OF STAINED MONOLAYERS

If cell growth is too low or if there is not enough cells to harvest for cell counting, cells can be stained *in situ* in monolayers and counted directly under the microscope. Because this procedure is tedious and subject to high operator error, isotopic labelling or the estimation of total amount of DNA or protein is preferable.

INDIRECT METHODS

Indirect measurements of viability are based on parameters like metabolic activity, DNA/protein content, cell weight, etc. The most commonly used parameter is glucose utilization. Oxygen utilization, lactic or pyruvic acid production, or carbon dioxide production can also be used. When cells are growing logarithmically, there is a very close correlation between nutrient utilization and cell numbers. However, during other growth phases, utilization rates, caused by maintenance rather than growth, can give misleading results.

The measurements obtained can be expressed as a growth yield (Y) or specific utilization/respiration rate (Q):

$$\text{Growth yield}(Y) = \frac{\text{Change in biomass concentration}(dx)}{\text{Change in substrate concentration}(ds)}$$

$$\text{Specific utilization / respiration rate}(Q_A) = \frac{\text{Change in substrate concentration}(ds)}{\text{Time}(dt) \times \text{cell mass / numbers}(dx)}$$

For example, typical values of growth yields for glucose (10^6 cells/g) are 385 (MRC-5), 620 (Vero), and 500 (BHK).

CELL WEIGHT

Wet weight can be used if very large number of cells is involved. Wet weight can give erroneous result as the amount of adherent extracellular liquid may vary in different cell lines. Dry weight also is seldom used, because salt derived from the medium contributes to the weight of the unfixed cells, and fixed cells lose some of their low-molecular-weight intracellular constituents and lipids.

PROTEIN DETERMINATION

The protein content of the cells is widely used for estimating total cellular material. The amount of protein in solubilized cells can be estimated directly by measuring the absorbance at 280 nm, with minimal interference from nucleic acids and other constituents. The absorbance at 280 nm can be detected down to 100 μm of protein, or about 2×10^5 cells.

Colorimetric assay for proteins can also be used. Among these assays, the Bradford reaction with Coomassie blue is one of the most widely used.

DNA DETERMINATION

DNA content can be measured in conjugation with fluorescent dyes. This method gives a sensitivity of 10 ng/ml, but requires intact double-stranded DNA. DNA can also be measured by its absorbance at 260 nm, where 50 μg/ml has an optical density of 1.0. Because of the interference with other cellular constituents, the direct absorbance method is useful only for purified DNA.

CYTOMETRIC ASSAY

In situ labelling with fluorescent dye or fluorescent antibody conjugate (FISH) can measure the amount of enzyme, DNA, RNA, protein or other cellular constituents *in situ* with a CCD camera (Charge-coupled device is an image sensor in digital photography). This process allows qualitative as well as quantitative analysis to be made, but is slow if large number of cells are to be scanned.

Flow cytometry can be used to measure a cell suspension of up to 1×10^7 cells. It can measure multiple cellular constituents and activities and enables to correlate these measurements with other cellular parameters, such as cell size, lineage, DNA content, or viability.

GROWTH KINETICS

The standard format of a culture cycle beginning with a lag phase, proceeding through the logarithmic phase to a stationary phase and finally to the decline and death of cells, is well-documented. Although cell growth usually implies increase in cell number, increase in cell mass can occur without any replication. The difference in mean cell mass between cell populations is considerable, as would be expected, but so is the variation within the same population.

Growth (increase in cell numbers or mass) can be defined in the following terms.

i. **Specific growth rate** (m) is the rate of growth per unit amount (weight/numbers) of biomass.

$$\mu = \left(\frac{1}{x}\right)\left(\frac{dx}{dt}\right)h^{-1}$$

where,

dx is the increase in cell mass,

dt is the time interval, and

x is the cell mass.

If the growth rate is constant (e.g., during logarithmic growth),

$$\ln x = \ln x_o + \mu t$$

where x_o is the biomass at time t_o.

ii. **Doubling time,** t_d, is the time for a population to double in number/mass.

$$t_d = \frac{\ln 2}{\mu} = \frac{0.693}{\mu}$$

iii. **Degree of multiplication, (n) or number of doublings** is the number of times the inoculum has replicated.

$$n = 3.32 \log(x/x_o)$$

GROWTH CURVE

The growth curve is a semilog plot of the increase in cell concentration (left axis) and cell density (right axis) after subculture. For plotting, replicate flask cultures are sampled daily and plotted against x-axis. Growth curve will be an elongated sigmoidal curve and exponential phase is represented by a straight line (Figure 8.3). Population doubling time (PDT) can be calculated from the middle of this straight line. The time at the intercept of the line extrapolated from the exponential phase with the seeding concentration is the lag time, and **the saturation density** is found at the plateau at the top end of the curve.

The log and plateau phases give vital information about the cell line. The measurement of PDT is used to quantify the response of the cells to different inhibitory or stimulatory culture conditions, such as variations in nutrient concentrations, hormonal effects, or toxic drugs. It is also a good monitor of the culture during serial passage and enables the calculations of the cell yields and the dilution factor required at the subculture. Growth curves are particularly useful for the determination of the saturation density in bioreactors. The PDT should not confuse with the cell cycle or generation time. PDT is an average figure that applies to the whole population and it describes the wide range of division rates taking place in the culture including non-dividing cells. The cell cycle time or generation time is measured from one point in the cell cycle until the same point is reached again, and refers to only the dividing cells in the culture. PDT varies from 12–15 h in rapidly growing cancer cell lines to 24–36 h in many adherent continuous cell lines and up to 60–70 h in slow-growing finite cell lines.

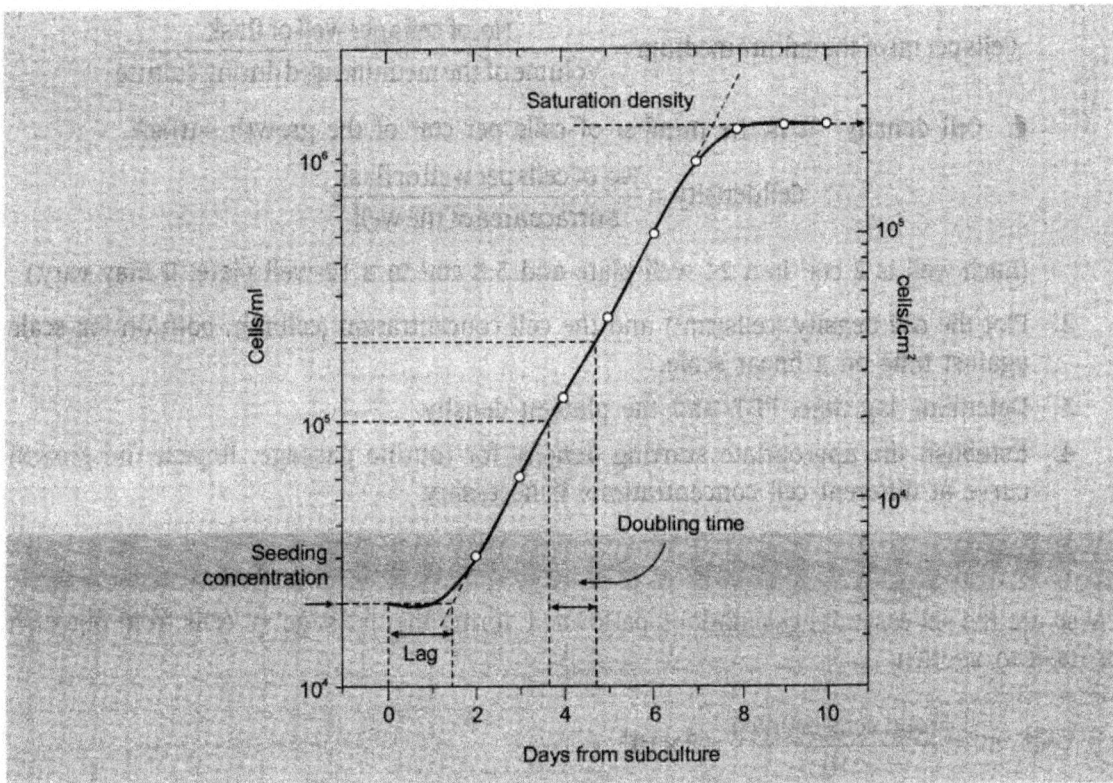

Figure 8.3 Growth curve. Growth kinetics of an adherent cell line showing various phases of cell growth

ANALYSIS OF GROWTH CURVE

A new growth cycle begins each time the culture is subcultured. A cell concentration of 2×10^4 cells/ml should be chosen for a rapidly growing line and 1×10^5 cells/ml for a slower growing cell line. Repeating the growth curve with higher or lower seeding concentrations should then allow the correct seeding concentrations and subculture interval to be established. Growth curve of monolayer culture can be prepared by the following steps.

1. Calculate the number of cells per well, per ml of culture medium and per cm² of available growth surface in the well as follows:

   ◗ **Primary count** This is the count obtained from the haemocytometer or the electronic counter and is the number of cells/ml of trypsinate.

   ◗ **Cells per flask/well** If 1 ml of trypsin is used, primary count is the same as the number of cells per flask/well.

   ◗ **Cell concentration** This is given by the following formula.

$$\text{Cells per ml of the culture medium} = \frac{\text{No. of cells per well or flask}}{\text{Volume of the medium used during culture}}$$

✦   **Cell density**   It is the number of cells per cm^2 of the growth surface.

$$\text{Cell density} = \frac{\text{No. of cells per well or flask}}{\text{Surface area of the well}}$$

(Each well is 2 cm^2 in a 24-well plate and 3.8 cm^2 in a 12-well plate. It may vary.)

2.   Plot the cell density (cells/cm^2) and the cell concentration (cell/ml), both on log scale against time on a linear scale.

3.   Determine lag time, PDT, and the plateau density.

4.   Establish the appropriate starting density for routine passage. Repeat the growth curve at different cell concentrations if necessary.

The Six Essential Calculations

These are the six essential calculations performed during the 'passage' of cells from one dish or flask to another.

1. $\text{cells / ml} = \dfrac{\text{Total cells counted}}{10} \times 2 \times 10^4$

where the cells are counted using haemocytometer, 2 is the dilution factor & 10^4 constant

2.   Total number of cells in dish = cells/ml $\times$ # ml media used to resuspend cells

(This information is useful when seeding a *large number of cells for multiple plate seedings*)

3. $\text{% Vialibility} = \dfrac{\text{No. of living cells}}{\text{Total no. of cells}} \times 100$

4.   Viable cells/ml =   Viability $\times$ total cells/ml

5. $\text{ml cell suspension needed} = \dfrac{\text{Total cells needed}}{\text{Viable cells / ml}}$

6.   Volume of fresh media = Total volume flask will hold – Volume of cell suspension added to dish

Total cells needed (as in step 4) is determined by

the size of the flask,

the size of the cells,

the doubling time of the cells and

the number of days you will be growing the cells before the next split

1. What are the direct and indirect methods used for quantitation of cells?

2. Write short notes on:

   i. Growth kinetics

   ii. Growth curve

9

CYTOTOXICITY ASSAYS

The total cell number, be it in a tissue *in vivo* or in a culture dish *in vitro*, is a balance between the rates of cell growth or mitosis, and cell death. The ability to differentiate between live and dead cells is therefore important. Many cell culture experiments carried out *in vitro* are solely for the purpose of determining the potential cytotoxicity of the compound being studied, usually a drug, a cosmetic or a food additive. Conventionally the toxicity is measured *in vivo* in experimental animals. *In vivo* toxicity is a complex event involving many physiological effects whose interpretation will be difficult. Moreover many countries have introduced new legislations against use of experimental animals especially on a large scale. So now most of the assays determine effects at a cellular level and are called **cytotoxicity assays**.

A cytotoxic assay should be simple, cheap, easily quantified and reproducible. Cell growth is the main criterion measured in most of the cytotoxicity assays. The net change in population size (through a growth curve), or a change in cell mass (total protein or DNA) or metabolic activity (DNA, RNA synthesis, MTT assay) are also measured in these assays.

NATURE OF THE ASSAY

The choice of the assay will depend on the agent under study, the nature of the response and the particular target cell. Assays can be divided into five major classes:

1. *Viability assays*   An immediate or short-term response, such as an alteration in membrane permeability or a perturbation of a particular metabolic pathway correlated with cell proliferation or survival.

2. *Survival assays*   The long-term retention of a self-renewal capacity (5–10 generations or more).

3. *Metabolic assays*   Assays, usually based on microtitrations, of intermediate duration that can either measure a metabolic response (e.g., dehydrogenase activity, DNA, RNA, or

protein synthesis) at the time of or shortly after exposure. Making the measurements after two or three population doubling after exposure is more likely to reflect cell growth and may correlate with cell survival.

4. *Calorimetric assays* These assays form colour development due to reaction of metabolites and formation of an end compound that quantifies the cell number. They are usually called as microtitration assays.

5. *Transformation assays* Survival in an altered state (e.g., a state expressing genetic mutation, alteration in growth control, or malignant transformation).

6. *Irritancy assays* Irritancy is a term analogous to inflammation, allergy, or irritation *in vivo*; as yet difficult to model *in vitro* but may be possible to assay by monitoring cytokine release in organotypic cultures.

VIABILITY ASSAYS

Viability assays are used to measure the proportion of viable cells after a potentially traumatic procedure, such as primary disaggregation, cell separation, or cryostorage.

VIABILITY TESTS BASED ON MEMBRANE INTEGRITY

This is the commonest measurement of cell viability at the time of assay. It will give an estimate of instantaneous damage (e.g., by cell freezing and thawing), or progressive damage over a few hours. Beyond this, quantitation may be difficult due to loss of dead cells by detachment and autolysis. These assays are of particular importance for toxic agents which exert their primary effect on membrane integrity. Important viability tests are described below.

i. *^{51}Chromium release* Labelling cells with ^{51}Cr results in covalent binding of chromate to basic amino acids of intracellular proteins. These labelled proteins leak out of the cell when the membrane is damaged, at a rate which is proportional to the amount of damage. The method is used in immunological studies for determining cytotoxic T-cell activity against target cells. Natural leakage of ^{51}Cr from undamaged cells may be high, and therefore the time period over which the assay can be used is restricted to approximately 4 hours.

ii. *Enzyme release assays* Enzyme release assays for measuring membrane integrity have been developed and these have overcome some of the problems of the ^{51}Cr release assay. Different enzymes have been used, though LDH has been found to be generally useful, since it is released by a range of cell types. The assay also has application in the wider context of toxicity testing, and has been used to investigate hepatotoxicity.

iii. *Dye exclusion* Most viability tests rely on a breakdown in membrane integrity measured by one of the following.

1. the uptake of a dye to which the cell is normally impermeable, e.g., trypan blue, erythrosin, or naphthalene black.

2. the release of a dye normally taken up by and retained by viable cells, e.g., diacetyl flourescein or neutral red.

However this effect is immediate and does not always predict ultimate survival. Furthermore dye exclusion tends to overestimate, e.g., 90% of the cells thawed from liquid nitrogen storage may exclude trypan blue, but only 60% prove to be capable of attachment 24 hours later.

Viability dyes used to determine membrane integrity include trypan blue, eosin Y, naphthalene black, nigrosin (green), erythrosin B, and fast green. Staining for viability assessment is more suited to suspension cultures than to monolayers, because dead cells detach from the monolayer and are therefore lost from the assay. A major disadvantage may be the failure of reproductively dead cells to take up dye, as demonstrated when cells with impaired clonogenicity showed 100% viability according to dye exclusion. The method has been renovated, however, and technical innovations introduced, which attempt to circumvent some of the problems commonly associated with such assays.

iv. *Fluorescent dyes*   There are a number of fluorescent probes now available for measuring membrane integrity.

- Calcein-AM—a membrane-permeant esterase substrate; cleaved by esterase in living cells to yield calcein, which fluoresces green in the cytoplasm.
- Ethidium-homodimer—a high-affinity red fluorescent nucleic acid stain which penetrates membrane of dead but not live cells.
- Propidium iodide—a red fluorescent nucleic acid stain which penetrates membrane of dead but not live cells.

v. *Neutral red uptake*   Living cells take up neutral red, 40 µg per ml. in the culture medium and retain it in lysosomes. But neutral red is not retained by non-viable cells. Uptake of neutral red can be quantified by fixing cells in formaldehyde and solubilizing the stain in acetic ethanol, allowing the plate to be read on an ELISA plate reader at 570 nm. Neutral red tends to precipitate, so the medium with stain is usually incubated overnight and centrifuged before use. This assay does not measure the total number of cells, but it does not show a reduction in the absorbance related to loss of viable cells, and is readily automated.

SURVIVAL ASSAYS

Although short-term tests are convenient and usually are quick and easy to perform, they reveal only cells that are dead (i.e., permeable) at the time of the assay. Frequently, cells subjected to toxic influence show an effect several hours or even days later. In these cases short-term toxicity as are not applicable and long-term survival assays are best suited. Survival implies the retention of regenerative capacity and is usually measured by **plating efficiency**. Plating efficiency measures survival by proliferative capacity for several cell generations, provided that the plate has a high-enough efficiency that the colonies can be considered representative of the entire cell population. Although not ideal, a plating efficiency of over 10% is usually acceptable.

PLATING EFFICIENCY

Colony formation at low cell density, or plating efficiency, is the preferred method for analysing cell proliferation and survival. This technique reveals differences in the growth rate within a population and distinguishes between alterations in the growth rate (colony size) and cell survival (colony number). It should be remembered, however that cells may grow differently in isolated colonies at low densities. In this situation fewer cells will survive, even under ideal conditions, and all cell interactions are lost until the colony starts to form. Heterogeneity in clonal growth rate reflects differences in the capacity for cell proliferation between lineages within a population, but these differences are not necessarily expressed in an interacting monolayer at higher densities, when cell communications are possible. When cells are plated out as a single cell suspension at low cell densities (2–25 cells/cm^2), they grow as discrete colonies. The number of these colonies can be used to express the plating efficiency:

$$\frac{\text{No. of colonies formed}}{\text{No. of cells seeded}} \times 100 = \text{Plating efficiency}$$

If it can be confirmed that each colony grew from a single cell, then the term becomes the **cloning efficiency**. It is also called **clonogenic assay**. Measurements of the plating efficiency are derived by counting the number of colonies over a certain size (usually around 50 cells). Growing from a low inoculum of cells, and this term should be used for the recovery of adherent cells after seeding at higher cell densities. Survival at higher densities is more properly referred to as the **seeding efficiency**.

$$\frac{\text{No. of cells attached}}{\text{No. of cells seeded}} \times 100 = \text{Seeding efficiency}$$

It should be measured at a time when the maximum number of cells has attached, but before mitosis starts.

METABOLIC ASSAYS

RESPIRATION AND GLYCOLYSIS

Drug-induced changes in respiration (oxygen utilization) and glycolysis (carbon dioxide production) have been measured using Warburg manometry. Another method is determination of dehydrogenase activity by incorporating methylene blue into agar-containing drug-treated cells, cell death being indicated by non-reduction of the dye. The latter method has the disadvantage of being non-quantitative, whilst the former, although quantitative, has not been widely adopted because the technical manipulations involved are extensive and unsuited for multiple screening. The more direct approach of monitoring pH changes in cultures containing an appropriate pH indicator has also been described.

RADIOISOTOPE INCORPORATION

Measurement of the incorporation of radiolabelled metabolites is a frequently used end-point for cytotoxicity assays of intermediate and short-term duration.

i. *Nucleotides* Measurement of [^{3}H] thymidine incorporation into DNA and [^{3}H] uridine incorporation into RNA are commonly used methods of quantitation of drug cytotoxicity. In short-term assays, which do not include a recovery period, there are a number of disadvantages, all of which relate to a failure of [^{3}H] thymidine incorporation to reflect the true DNA synthetic capacity of the cell. These are as follows:

- Changes may relate to changes in the size of the intracellular nucleotide pools rather than changes in DNA synthesis.
- Some drugs such as 5-fluorouracil and methotrexate which inhibit pyrimidine biosynthesis (de novo pathway) cause increased uptake of exogenous [^{3}H] thymidine due to a transfer to the 'salvage' pathway, which utilizes preformed pyrimidines.
- Continuation of DNA synthesis in the absence of [^{3}H] thymidine incorporation can occur.

The low labelling index of human tumours with resultant low levels of nucleotide incorporation in short-term assays necessitates the use of high cell densities, which can restrict the number of drugs and range of concentrations tested when cell numbers are limited. Two 'hybrid' techniques have been reported, which combine the advantages of the soft agar culture system with the facilitated quantitation offered by the use of radioisotopes. Both assays are of intermediate duration (about four days) and use [^{3}H] thymidine incorporation into DNA as an end-point. In one method the cells are grown in liquid suspension over soft agar, whilst in the other the cells are incorporated in the soft agar. Given that a homogeneous cell population is available, [^{3}H] nucleotide incorporation can be used after an appropriate recovery period to measure survival or, in the presence of drug, to measure an antimetabolic effect, but with the reservations expressed above.

ii. [^{125}I]*Iododeoxyuridine* ([^{125}I]*Udr*) [^{125}I]Udr is a specific, stable label for newly synthesized DNA which is minimally reutilized and can therefore be used over a 24 hours period to measure the rate of DNA synthesis; quantitation is facilitated because the isotope is a gamma emitter. Disadvantages include its variable toxicity to different cell populations, which therefore means that more cells are required because [^{125}I]Udr must be used at low concentrations.

iii. *[^{32}P]Phosphate (^{32}P)* The rate of release of ^{32}P into the medium from pre-labelled cells is a function of the cell type and is increased in damaged cells. This has been used as a measure of drug efficacy. The incorporation of ^{32}P into nucleotides has also been used as an index of drug cytotoxicity. Neither method has been routinely adopted.

iv. *[^{14}C]Glucose* Glucose incorporation is used as a cytotoxicity end-point because it is a precursor which is common to a number of biochemical pathways. The method has not been widely used.

v. *[^{3}H]Amino acids* Protein synthesis is an essential metabolic process without which the cell will not survive, and incorporation of amino acids into proteins has been used as an index of cytotoxicity. The most extensive studies have utilized monolayers of cells growing in microtitration plates, using either incorporation of [^{3}H]leucine measured by liquid scintillation counting, or [^{35}S]methionine incorporation, measured using autofluorography.

vi. *^{45}Calcium (^{45}Ca)* Unrelated compounds may produce alterations in the permeability of cell membranes to calcium, such that increased calcium uptake results. Measurements of ^{45}Ca uptake can therefore be used.

TOTAL PROTEIN CONTENT

Protein content determination is a relatively simple method for estimating cell number. It is particularly suited to monolayer cultures, and has the advantage that washed; fixed samples can be stored, refrigerated for some time before analysis without impairment of results, facilitating large-scale screening. Overestimation of cell number may arise with drugs which inhibit replication without inhibiting protein synthesis (e.g., BrdU, methotrexate). Assessment of cytotoxicity requires the demonstration of an alteration in the accumulation of protein per culture over time, preferably at several points, or at one point after prolonged drug exposure and recovery.

COLORIMETRIC ASSAYS (MICROTITRATION ASSAYS)

The advent of sophisticated microplate readers which allow rapid quantitation of colorimetric assays has paralleled the development of a variety of assays which use some form of colour development as an end-point for quantitating cell number. These assays are generally called microtitration assays. These include methods which reflect:

- protein content (methylene blue, Coomassie blue, Kenacid blue, sulphorhodamine B, Bichinoninic acid)
- DNA content or DNA synthesis (BrdU uptake)

♦ lysosome and Golgi body activity (neutral red)

♦ enzyme activity (hexosaminidase, mitochondrial succinate dehydrogenase)

Linear relationships between end-point and cell number have been demonstrated for all these methods. Discrimination between live and dead cells in monolayer assays is not relevant, since dead cells will usually detach, given sufficient time. In suspension cultures, this aspect is relevant, however, and needs to be considered when choosing an appropriate assay. Methods which make use of fluorescent dyes have increased in the past five years and bioluminescent assays have also been developed for measuring ATP levels.

i. *Protein content* Several methods are available for measuring the protein content of cell monolayers. These include the use of the Folin-Ciocalteau reagent according to the method of Lowry, and amido black. Several new techniques for colorimetric determination of protein content are also available. These include methylene blue, sulphorhodamine B, Kenacid blue, Coomassie blue G-250, and Bichinoninic acid (BCA).

ii. *DNA content* DNA content may be measured in microtitration plates by staining with dyes whose fluorescence is enhanced by intercalation at AT-specific sites on the chromatin, such as Hoechst 33342 and 2-diamidinophenylindole (DAPI). DNA synthesis may also be determined using bromodeoxyuridine (BrdU). The amount of BrdU incorporated is detected immunohistochemically using a monoclonal antibody to BrdU, and the binding may be quantitated using appropriate conjugates and chromogenic substrates.

iii. *Lysosomal and Golgi body activity* The uptake of neutral red by lysosomes and Golgi bodies has been used to quantitate cell number. The stain appears to be specific for viable cells, but the main limitation of the method is the difference in uptake between cell types. Thus, some cell types, such as activated macrophages and fibroblasts, take up large amounts very rapidly whereas others, such as lymphocytes, show negligible staining.

iv. *Tetrazolium dye reduction (MTT assay)* MTT assay is a standard colorimetric assay for measuring cellular proliferation (cell growth). Yellow MTT [3-(4,5-Dimethylthiazol-2-yl)-2,5-diphenyltetrazolium bromide, a tetrazole] is reduced to purple formazan in the mitochondria of living cells. A solubilization solution (usually either dimethyl sulphoxide or a solution of the detergent sodium dodecyl sulphate in dilute hydrochloric acid) is added to dissolve the insoluble purple formazan product into a coloured solution. The absorbance of this coloured solution can be quantified by measuring at a certain wavelength (usually between 500 and 600 nm) by a spectrophotometer.

This reduction takes place only when mitochondrial reductase enzymes are active, and therefore conversion is directly related to the number of viable (living) cells. When the amount of purple formazan produced by cells treated with an agent is compared with the amount of formazan produced by untreated control cells, the effectiveness of the agent in causing death of cells can be deduced, through the production of a dose–response curve.

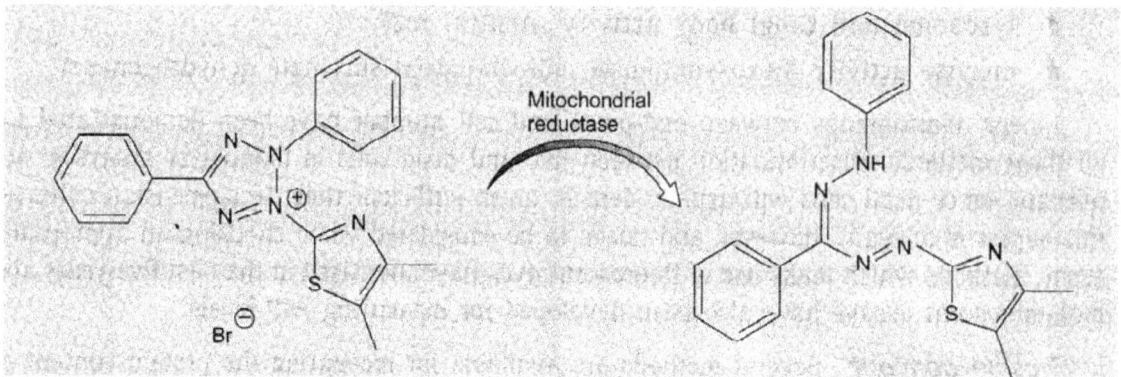

The method was first described in 1983 as a rapid colorimetric method for immunological studies, and modifications for this application have been described. The technique is particularly useful for assaying cell suspensions because of its specificity for living cells. One disadvantage is the need to use unfixed cells, which may impose time restraints. The potential of the technique for drug sensitivity testing of human tumours was recognized.

v. *Luminescence-based cell viability testing* This is a viability assay, which combines the lytic reagent with luciferase/luciferin and allows sensitive detection of the end metabolic product in a single step. The assay system utilizes a reagent formulation containing a highly stabilized luciferase, to extract ATP from the cells and support a stable "glow-type" luminescent signal. Levels of ATP in a cell population provide a sensitive indication of cell viability. The sensitivity range has been reported as 20 cells/ml to 2×10^7 cells/ml. A number of commercial kits are available for measuring ATP by luminescence.

TRANSFORMATION ASSAY

Transformation of normal cells into neoplastic cells occurs through a series of genetic alterations, yielding a cell population capable of proliferation independently that normally restrain their growth. Traditionally, the soft agar colony formation assay has been used to monitor cell transformation and anchorage-dependent growth, with manual counting of proliferated cells after 3–4 weeks of cell growth. Transformation detection assay is an anchorage-independent growth assay in soft agar and a stringent assay for detecting malignant transformation of cells. Cells pre-treated with carcinogens or carcinogen inhibitors are cultured with appropriate controls in soft agar medium for 21–28 days. Following this incubation period, formed colonies can either be analysed morphologically using cell stain solution or the cells are quantified. Neoplastic cell transformation results from the accumulation of genetic mutations that permit uncontrolled proliferation, increased growth potential, alterations in cell surface, karyotypic abnormalities, morphological and biochemical deviations and other attributes conferring the ability to invade, metastasize and kill.

MELISA (Memory lymphocyte immunostimulation assay) test is modification of the transformation assay which detects Type IV hypersensitivity (allergy) to metals, chemicals, environmental toxins and moulds. The test will not measure the amounts of a harmful substance but measures the allergic state.

IRRITANCY ASSAY

The common irritancy test is the "Draize test". Initially used for testing cosmetics, the procedure involves applying 0.5 ml or 0.5 g of a test substance to the eye or skin of a restrained, conscious animal and leaving it for four hours. The animals are observed for up to 14 days, for signs of redness, swelling, discharge, ulceration, haemorrhaging, cloudiness or blindness in the tested eyes. The test subject is commonly an albino rabbit, though other species are used too, including dogs. The animals are euthanized after testing if the test renders irreversible damage to the eyes or skin. Animals may be re-used for testing purposes if the product tested causes no permanent damage. Animals are typically re-used after a "wash out" period during which all traces of the tested product are allowed to disperse from the test site. Such tests are subjected to ethical clearances. An alternative to such testing is the *in vitro* irritancy testing in cell monolayers to identify the cytotoxicity and apoptic tendency of the cells. Tests are performed both in primary and established cell lines depicting the same origin of the cells. For example, a dermal irritancy test could be performed in keratinocyte cells.

APOPTOSIS MEASUREMENTS

Many anticancer drugs kill cells by apoptosis, and measurement of apoptosis is therefore important in the evaluation of cytotoxicity. Apart from morphological criteria, apoptosis can be determined in a number of ways including:

DNA laddering or Comet assay   Single cell gel (SCG) electrophoresis or 'Comet assay' is a rapid and very sensitive fluorescent microscopic method to examine DNA damage and repair at individual cell level.   This assay has critical applications in fields of toxicology ranging from aging and clinical investigations to genetic toxicology and molecular epidemiology.

TUNEL assay   Also known as TRITC-mediated dUTP nick end-labelling assay or terminal deoxynucleotidyl transferase assay, this assay demonstrates the late stage of apoptosis, or programmed cell death, by detecting the fragmentation of nuclear chromatin. The TUNEL assay detects apoptosis-induced DNA fragmentation through a quantitative fluorescence assay. Terminal deoxynucleotidyl transferase (TdT) catalyses the incorporation of bromodeoxyuridine (BrdU) residues into the fragmenting nuclear DNA at the 3´-hydroxyl ends by nicked end labelling. A fluorescent dye (TRITC-tetramethylrhodamine isothiocyanate) conjugated to anti-BrdU antibody can be used for labelling the 3´-hydroxyl ends for detection by PCA system.

Caspase assays   The caspases consist of a group of aspartic-acid-specific cysteine proteases, which are activated during apoptosis. These unique proteases, which are synthesized as zymogens, are involved in the initiation and execution of apoptosis once activated by proteolytic cleavage. Mammalian caspases may be grouped by function: cytokine activation includes caspases 1, 4, 5, 13; apoptosis initiation includes caspases 2, 8, 9, 10; and apoptosis execution utilizes caspases 3, 6, 7. Caspase assays are based on the measurement of zymogen processing to an active enzyme and proteolytic activity.

The caspase assays measure caspase (involved in apoptotic signalling) activities using a luminogenic caspase substrate and a luciferase in a reagent optimized for specific caspase activity, luciferase activity and cell lysis. Adding the single caspase reagent in an "add-mix-measure" format results in cell lysis, followed by caspase cleavage of the substrate. This cleavage liberates free aminoluciferin, which is consumed by the luciferase, generating a "glow-type" luminescent signal. The signal is proportional to caspase activity present. Currently caspase assays are available to measure caspase-3/7, caspase-8 or caspase-9 activity. These assays can be multiplexed with cell viability and cytotoxicity assays.

REVIEW QUESTIONS

1. Name the different cytotoxic assays.

2. What do you understand by viability assays in cytotoxicity test? Explain the different types of viability tests.

3. What are the different types of metabolic assays?

4. What do you understand by apoptosis? Explain the different methods used for its assay.

10

SCALING UP OF CULTURES

Scale-up of cultures primarily involves an increase in the volume of the culture medium. Bioreactors of capacities ranging from 100 to 1000 L are required for a range of production cells—from the laboratory scale requiring 1×10^{11} cells to the semi-industrial scale requiring about 1×10^{12} cells. Full-scale industrial production uses 5000–20,000 L bioreactors. The term **stirrer culture** or **spinner culture** is used for the simple growth of cells in the laboratory range of equipment. The term **fermenter** is generally used for laboratory-level cultures with a capacity ranging from 50 to100 L. The **bioreactor** is termed for variously designed culture equipment for scale-up and large-industrial scale production.

GENERAL CONSIDERATIONS FOR FERMENTING MAMMALIAN CELLS

OXYGEN LIMITATION

Oxygen limitation is usually the first factor to be overcome in culture scale-up. Oxygen is only sparingly soluble in culture media (7.6 µg/ml) and the oxygen utilization rates by cells have a mean value of 6 µg/10^6 cells/h. A typical culture of 2×10^6 cells/ml would, therefore, deplete the oxygen content of the medium (7.6 µg/ml) in less than 1 h. It is necessary to supply oxygen to the medium throughout the life of the culture. Thus scale-up of animal cell cultures is very much dependent upon the ability to supply sufficient oxygen without causing cell damage. This becomes a problem in conventional stirred cultures at volumes above 10 litres. Agitation of the medium by sparging with CO_2 and air is necessary when the depth exceeds 5 mm for adequate gas exchange.

Sparging This is the bubbling of gas through the culture, and is a very efficient means of effecting oxygen transfer. However, it may be damaging to animal cells due to the effect of high surface energy of the bubble on the cell membrane. This damaging effect can be minimized by using large air bubbles (which have lower surface energies than small bubbles), by using a very low gassing rate (e.g., 5 ml/1 min), and by adding surfactants. When sparging is used, efficiency of oxygenation is increased by using a culture vessel with a large height/diameter ratio. This creates a higher pressure at the base of the reactor, which increases oxygen solubility.

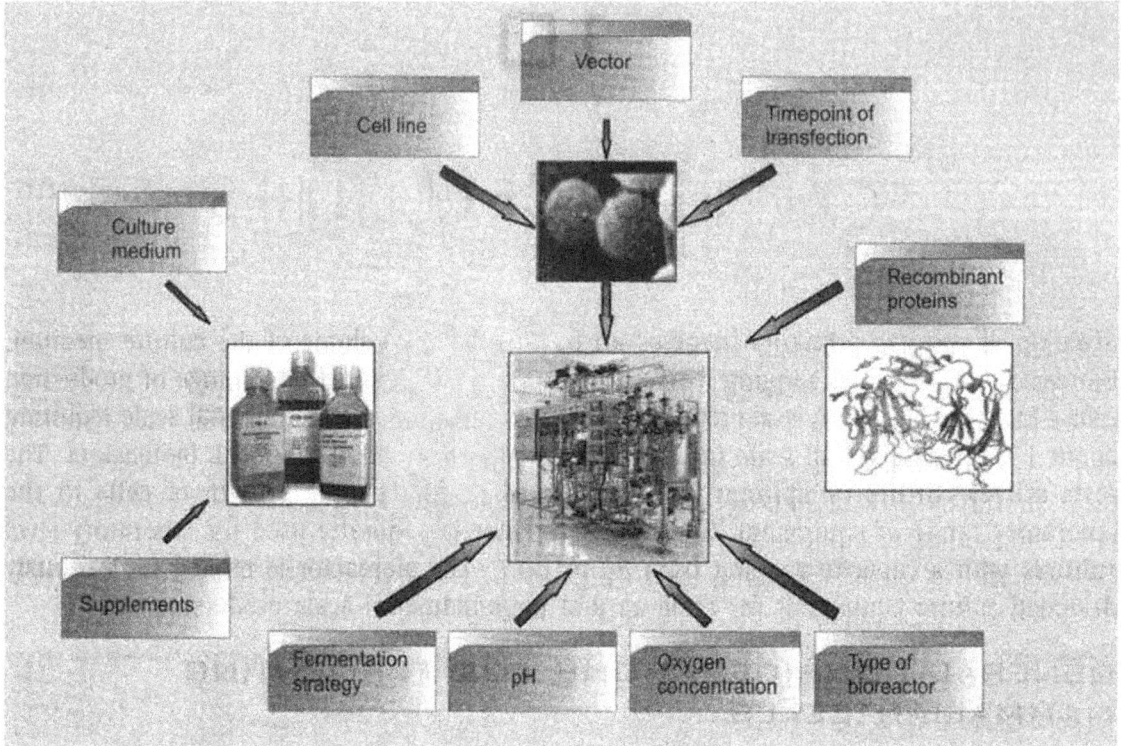

Figure 10.1 Scaling-up: An overview (From *Culture of Animal Cells* by R.I. Freshney)

Membrane diffusion Silicone tubing is very permeable to gases, and if long lengths of thin-walled tubings can be arranged in the culture vessel, sufficient diffusion of oxygen into the culture can be obtained. However, a lot of tubing is required (e.g., 30 m of 2.5 cm tubing for a 1000 litre culture). This method is expensive and inconvenient to use.

Medium perfusion A closed-loop perfusion system continuously takes medium from the culture, passes it through an oxygenation chamber, and returns it to the culture. This method has many advantages if the medium can be conveniently separated from the cells for perfusion through the loop. The medium in the chamber can be vigorously sparged to ensure oxygen saturation, and addition of other substances such as sodium hydroxide for pH control, which would damage the cells if put directly into the culture. This method is used in glass-bead systems and has proved particularly effective in microcarrier systems, where specially modified spin filters can be used.

Environmental supply The dissolved oxygen concentration can be increased by increasing the headspace pO_2 (from atmospheric 21% to any value, using oxygen and nitrogen mixtures) and by raising the pressure of the culture by 100 kPa (which increases the solubility of oxygen and its diffusion rate). These methods should be employed only when the culture is well-advanced, otherwise oxygen toxicity could occur. Finally, the geometry of the stirrer blade also affects the oxygen transfer rate.

MEDIUM AND NUTRIENTS

A given concentration of nutrients can support only a certain number of cells. Alternative nutrients can often be found by a cell when one becomes exhausted, but this is a bad practice because the growth rate is always reduced. Nutrients likely to be exhausted first are glutamine, because it spontaneously cyclizes to pyrrolidone carboxylic acid and is enzymically converted (by serum and cellular enzymes) to glutamic acid, leucine, and isoleucine. Human diploid cells are almost unique in utilizing cystine heavily. Hence the nutrients become growth-limiting before they become exhausted. As the concentration of amino acids falls, the cell finds it increasingly difficult to maintain sufficient intracellular pool levels.

Glucose is often another limiting factor as it is destructively utilized by cells and, rather than adding high concentrations at the beginning, it is more beneficial to supplement after two to three days. In order to maintain a culture, some additional feeding often has to be carried out either by complete, or partial media changes or by perfusion. Many cell types are either totally dependent upon certain additives or can perform optimally only when they are present. For many purposes it is highly desirable, or even essential, to reduce the serum level to 1% or below. In order to achieve this without a significant reduction in cell yield, various growth factors and hormones are added to the basal medium. The most common additives are insulin (5 mg/litre), transferrin (5–35 mg/litre), ethanolamine (20 µM), and selenium (5 µg/litre). Cell aggregation is often a problem in suspension cultures. Media lacking calcium and magnesium ions have been designed specifically for suspension cells because of the role of these ions in attachment. This problem has also been overcome by including very low levels of trypsin in the medium (2 µg/ml).

pH VARIATIONS

Ideally pH should be near 7.4 at the initiation of a culture and not fall below a value of 7.0 during the culture, although many hybridoma lines appear to prefer a pH of 7.0 or lower. A pH below 6.8 is usually inhibitory to cell growth. The factors that affect the pH stability of the medium include buffer capacity and type, headspace, and glucose concentration. The normal buffer system in tissue culture media is the carbon dioxide–bicarbonate system analogous to that in blood. This is a weak buffer system, in that it has a pKa well below the physiological optimum. It also requires the addition of carbon dioxide to the headspace above the medium to prevent the loss of carbon dioxide and an increase in hydroxyl ions.

The buffering capacity of the medium is increased by the phosphates present in the balanced salt solution (BSS). Medium intended to equilibrate with 5% carbon dioxide usually contains Earle's BSS (25 mM $NaHCO_3$) but an alternative is Hanks' BSS (4 mM $NaHCO_3$) for equilibration with air. Improved buffering and pH stability in media is possible by using a zwitterionic buffer, such as HEPES (10–20 mM), either in addition to, or instead of, bicarbonate.

The headspace volume in a closed culture is important because, in the initial stages of the culture, 5% carbon dioxide is needed to maintain a stable pH in the medium but, as the cells grow and generate carbon dioxide, it builds up in the headspace and this prevents it diffusing out of the medium. The result is an increase in weakly dissociated bicarbonate producing an excess of hydrogen ions in the medium and a fall in pH. Thus, a large headspace is required in closed cultures, typically tenfold greater than the medium volume (this volume is also needed to supply adequate oxygen). This generous headspace is not possible as cultures are scaled-up, and an open system with a continuous flow of air, supplied through one filter and extracted through another, is required.

The metabolism of glucose by cells results in the accumulation of pyruvic and lactic acids. Glucose is metabolized at a far greater rate than needed. Thus, glucose should never be included in media at concentrations above 2 g/litre, and it is better to supplement during the culture than to increase the initial concentration. An alternative is to substitute glucose by galactose or fructose as this significantly reduces the formation of lactic acid, but it usually results in a slower growth rate. These precautions delay the onset of a nonphysiological pH and are sufficient for small cultures. As scale-up increases, headspace volume and culture surface area in relation to the medium volume decrease. Also, many systems are developed in order to increase the surface area for cell attachment and cell density per unit volume. Thus pH problems occur far earlier in the culture cycle because carbon dioxide cannot escape as readily, and more cells means higher production of lactic acid and carbon dioxide. The answer is to carry out frequent medium changes or use perfusion, or have a pH control system. The basis of a pH control system is an autoclavable pH probe. This feeds a signal to the pH controller, which is converted to give a digital or analog display of the pH. This is a pH monitor system.

REDOX POTENTIAL

The oxidation–reduction potential (ORP), or redox potential, is a measure of the charge of the medium and is affected by the proportion of oxidative and reducing chemicals, the oxygen concentration, and the pH. When fresh medium is prepared and placed in the culture vessel, it takes time for the redox potential to equilibrate, a phenomenon known as **poising**. The optimum level for many cell lines is +75 mV, which corresponds to a dissolved pO_2 of 8–10%. If the redox potential is monitored by means of a redox electrode and pH meter (with mV display), an indication of how cell growth is progressing can be obtained. This is because the redox value falls during logarithmic growth and reaches a minimum value approximately 24 h before the onset of the stationary phase. This provides a useful guide to cell growth in cultures where cell sampling is not possible. It is also useful to be able to predict the end of the logarithmic growth phase so that medium changes, addition of virus, or product promoters can be given at the optimum time.

SUSPENSION-ADAPTING CELLS

If repeated culturing efforts with same cell line are required, suspension growth in spinners is probably optimal. Anchorage-dependent cells can be conditioned for suspension culture by a process called suspension adaptation. The purpose of suspension adaptation is to obtain cells that will grow as single cells unattached to a substrate. There are several approaches to reaching this goal:

- alter the medium to eliminate or diminish the ability of cells to attach to the glass or plastic substrate;
- select for cells that will not attach in the standard medium and substrate conditions; or
- select for cells that will grow in suspension in the standard media when the surfaces available have been treated to prevent attachment (these cells may still attach to surfaces treated for tissue culture).

The first approach has the disadvantage that the media devised to prevent cell attachment (generally with much reduced magnesium and calcium, e.g., Joclik's medium) are frequently sub-optimal for supporting the secretion of high titers of desired proteins. The second approach is adequate for production cell lines but is more difficult than the third approach, and the resulting cell lines are less flexible. The third approach is generally (but by no means always) rapid and it results in a line that can still be grown in an attached fashion if desired for further manipulation, such as cloning or transfection. The approach outlined below is designed to suspension-adapt cells in this third sense, with as little alteration in other cell properties as possible. The one exception to this rule is that we sometimes have chosen to suspension-adapt in a reduced serum (or serum-free), hormone-supplemented medium in order to obtain a line that will grow continuously in these conditions. In at least one case, this strategy also improved our ability to suspension-adapt the cells and to obtain a stable phenotype.

SCALE-UP OF SUSPENSION CULTURE

To support the high cell densities required for economic operation, the suspension must be agitated to ensure consistent and effective mixing so that all of the cells have access to nutrients and oxygen, and excess heat and exhaust gas are removed. The agitation method differentiates the two basic types of suspension bioreactors.

1. **Air-lift fermenter** where agitation is supplied by the movement of air through bioreactor.

2. **Stirred-tank fermenter** where mechanical impellers agitate the suspension.

For small volumes, it is possible to use alternatives such as shaker flasks, mixing bowls, and disposable-wave bioreactors. High cell densities in suspension cultures are achieved by consistent and efficient agitation to ensure that all the cells have access to the required nutrients, and excess heat and exhaust gas are removed. Irrespective of the agitation method, bioreactors are designed to avoid the introduction of any unwanted materials, or contaminants. This requires careful design of all connections, tubing, and the vessel itself to prevent areas that cannot be properly cleaned and sterilized. Bioreactors and associated pipework surfaces are usually constructed of stainless steel and have finish of high standards to reduce attached growth and to aid cleaning. Any entry or exit point from the vessel is a potential contamination route. The overall risk of contamination is often reduced by running the entire vessel slightly above atmospheric pressure, impeding the entry of external organisms. Cell cultures produce heat, and the larger the working volume, the greater the heat load. The vessel is therefore cooled during the fermentation process to ensure that the process operating temperature remains at the optimum value. Cooling is often supplied by external cooling jackets or coils. Internal coils may also be used to allow more rapid cooling of vessels following sterilization. This may be of greatest benefit if the process requires the sterilization of growth media *in situ*.

STIRRED-TANK BIOREACTORS

In contrast to air-lift bioreactors, stirred-tank bioreactors use mechanical devices to agitate the growth medium. Impellers and baffles within the vessel cause turbulence and allow finer control of the oxygen and mass transfer processes. Stirred-tank systems are readily scaleable, and typically use a more squat configuration, with aspect ratios of approximately 3 : 1 or less. The internal configuration of the vessel has a direct impact on the efficiency of the mixing, and many different designs exist. The mixing is achieved using rotating impellers and fixed baffles attached to the inside of the vessel. Four to six baffles approximately one-tenth of the overall vessel diameter are commonly used. Stirred-tank reactors generally contain a series of impellers. A large number of different impeller designs exist and may be used in specific applications to generate the required agitation. The microorganism being grown determines the choice of impeller design. Impellers can be either a combination or single use of the two main types:

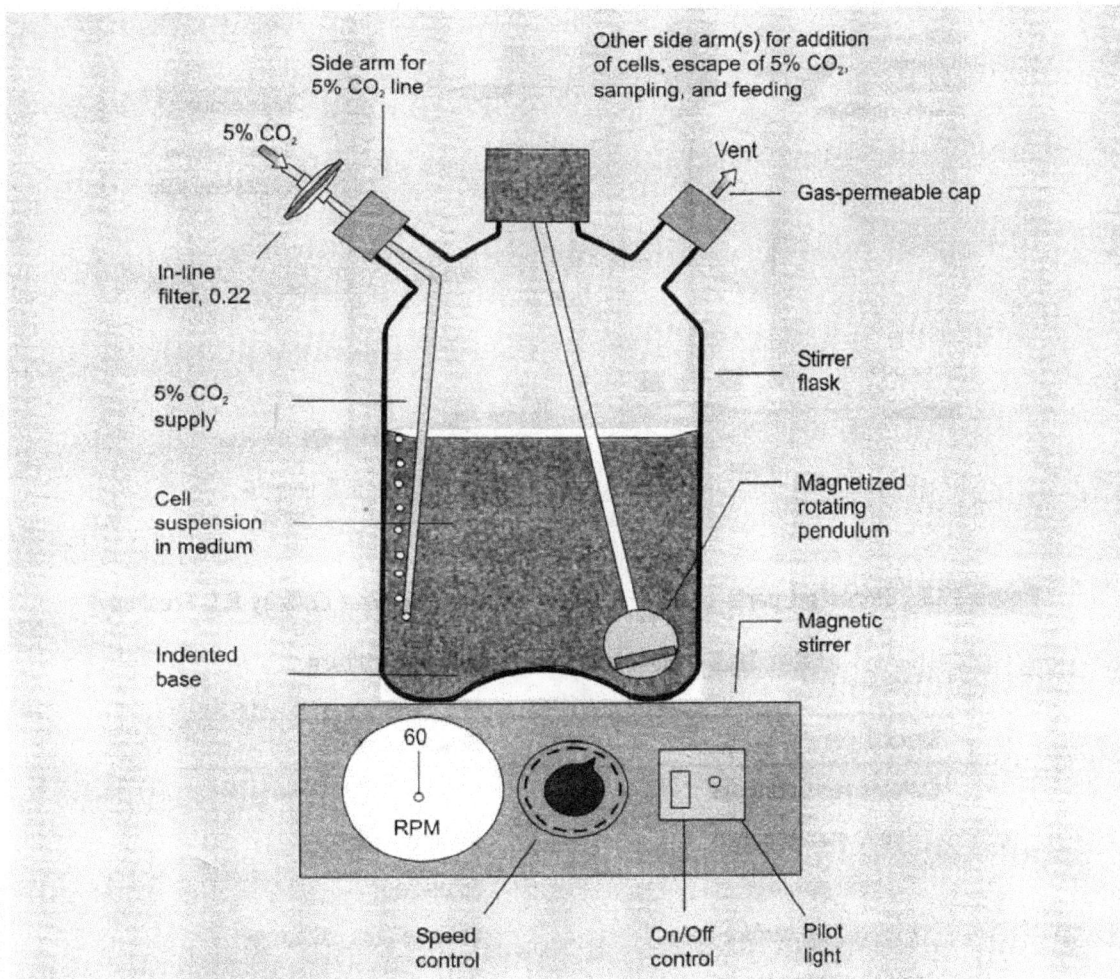

Figure 10.2 Stirrer culture. Diagram of a large stirrer flask suitable for volumes up to 8 L. (From *Culture of Animal Cells* by R.I. Freshney)

1. **Axial dispersion** designed very much like a ship propeller. This type of impeller moves the fluid towards the top of the vessel.
2. **Radial flow** that moves the fluid laterally around the tank.

The drive for the impeller system is supplied by an external motor and can be attached either to the top of the vessel or to the bottom. In both cases, the design of the impeller shaft seals is important because they must prevent the entry of any contaminant and in the case of a bottom-driven impeller arrangement, the leakage of the broth itself. Sealing methods often use condensate seals fed by clean process steam to ensure sterile operation. Organisms that are easily damaged must be grown in vessels using impellers and baffles that do not generate shear forces that destroy them.

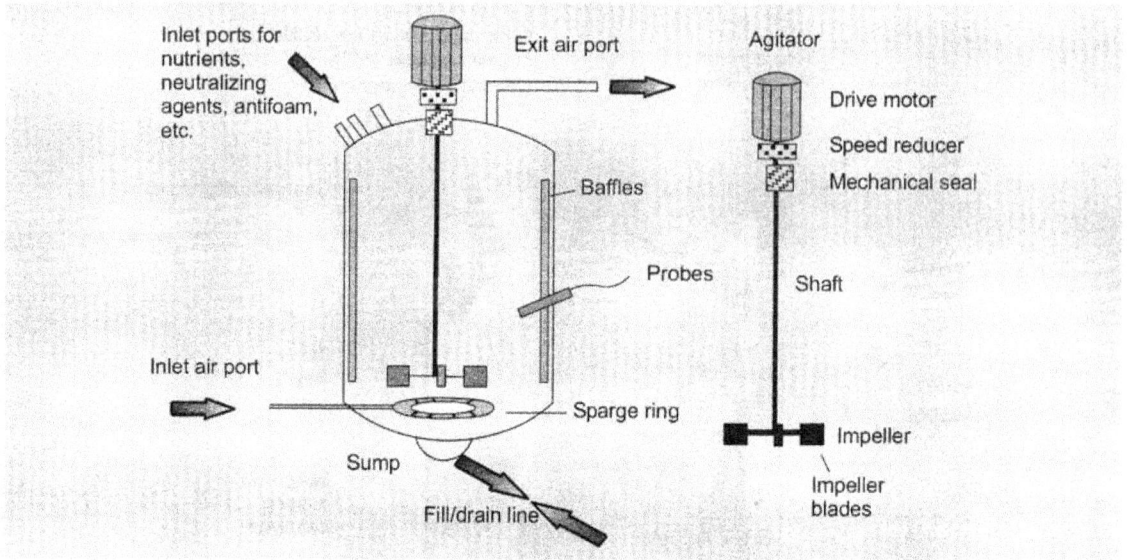

Figure 10.3   Important parts of an STR (From *Culture of Animal Cells* by R.I. Freshney)

Table 10.1   Monolayer vs. suspension culture

Monolayer	Suspension
Culture requirements	
Cyclic maintenance	Steady-state
Trypsin passage	Dilution
Limited by surface area	Volume (gas exchange)
Growth properties	
Contact inhibition	Homogeneous suspension
Cell interaction	
Diffusion boundary	
Useful for	
Cytology	Bulk production
Mitotic shake-off	Batch harvesting
in situ extractions	
Continuous product harvesting	
Applicable to	
Most cell types, including primaries	Only transformed cells

CELL DAMAGE IN ANIMAL CELL BIOREACTORS

The successful cultivation of animal cells requires that mass and heat transfer requirements be met. Agitation and aeration are thus critical considerations in the large-scale cultivation of animal cells. Unlike microbial cells, animal cells do not have cell walls and are protected from environmental forces only by their enriched cell membranes. Animal cells are therefore regarded as **shear-sensitive**. There are two major physical forces that can cause cell damage: shear forces and bubble energy.

SHEAR FORCES

Shear forces are created from fluctuating liquid velocities, which arise during turbulent mixing and are visualized as turbulent eddies. Shear forces increase with the level of turbulence and on the type of agitator used. Animal cells do not have cell walls and therefore can be damaged by shear.

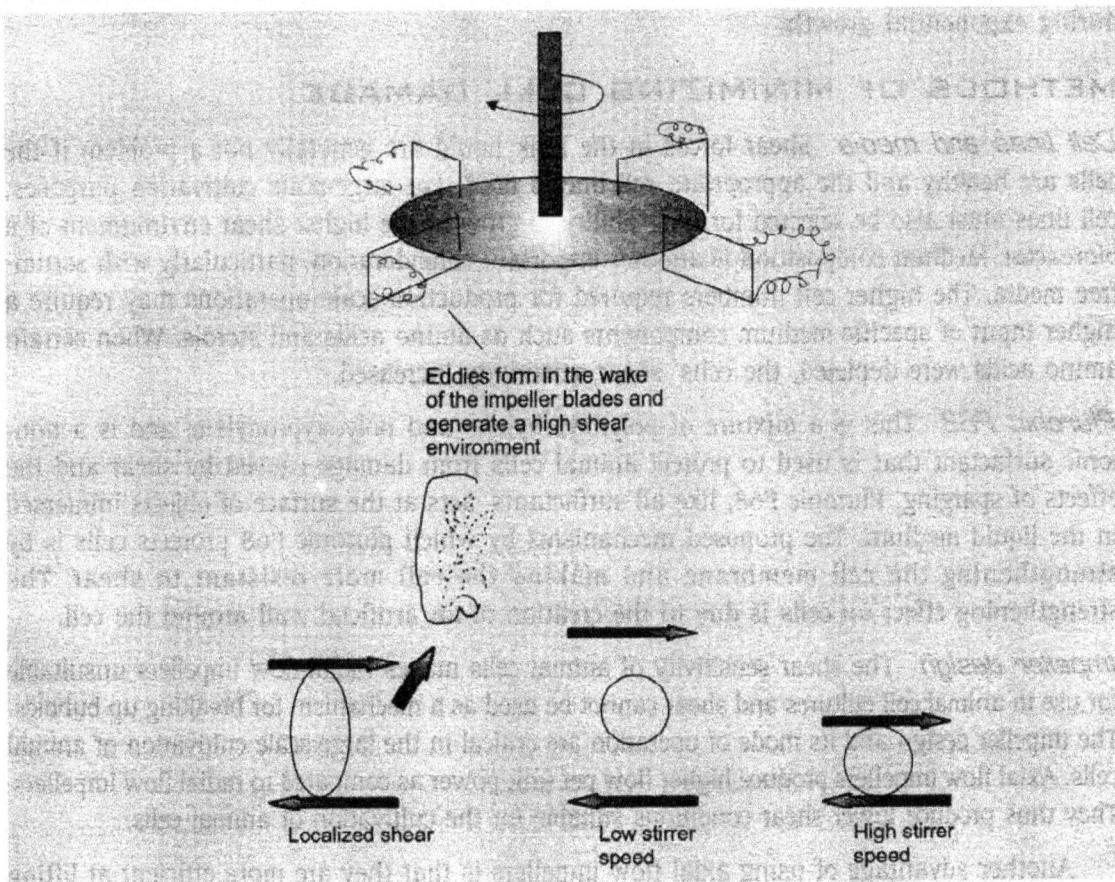

Figure 10.4 Causes of cell damage due to localized shear and shear stress (From *Culture of Animal Cells* by R.I. Freshney)

There are two "forms" of shear that we are concerned about when we consider the damage that shears can do to animal cells. These are

1. Localized shear which occurs around objects moving in the culture media, e.g., impellers and bubbles.
2. Shear in the bulk liquid arising from turbulence with the reactor.

Shear arising as a result of bubbles moving through the bulk liquid is not considered a major cause of cell damage. In sparged and baffled reactors, the cell damage is due to **turbulence,** i.e., the flow pattern is turbulent. In baffled reactors, as the stirrer speed increases, turbulent eddies will be formed in the bulk liquid; as the level of turbulence increases, the eddy size will decrease and the level of shear will increase. For this reason axial flow impellers are used in the culture of animal cells. It should also be noted that the sensitivity of animal cells to liquid shear forces varies with the cell line and age. Cells have been found to be more fragile during stationary and lag phases. Their robustness increases during exponential growth.

METHODS OF MINIMIZING CELL DAMAGE

Cell lines and media   Shear forces in the bulk liquid are generally not a problem if the cells are healthy and the appropriate cell line is used. For large-scale cultivation purposes, cell lines must also be selected for their ability to grow in the higher shear environment of a bioreactor. Medium composition is another important consideration, particularly with serum-free media. The higher cell numbers required for production-scale operations may require a higher input of specific medium components such as amino acids and sterols. When certain amino acids were depleted, the cells' shear sensitivity increased.

Pluronic F68   This is a mixture of polyoxyethylene and polyoxypropylene and is a non-ionic surfactant that is used to protect animal cells from damage caused by shear and the effects of sparging. Pluronic F68, like all surfactants, acts at the surface of objects immersed in the liquid medium. The proposed mechanisms by which pluronic F68 protects cells is by strengthening the cell membrane and making the cell more resistant to shear. The strengthening effect on cells is due to the creation of an artificial wall around the cell.

Impeller design   The shear sensitivity of animal cells makes radial-flow impellers unsuitable for use in animal cell cultures and shear cannot be used as a mechanism for breaking up bubbles. The impeller design and its mode of operation are critical in the large-scale cultivation of animal cells. Axial flow impellers produce higher flow per unit power as compared to radial flow impellers. They thus produce lower shear conditions suitable for the cultivation of animal cells.

Another advantage of using axial flow impellers is that they are more efficient at lifting cells from the base of the reactor. Axial flow impellers stirring at relatively low stirrer speeds are therefore widely used in the culture of animal cells.

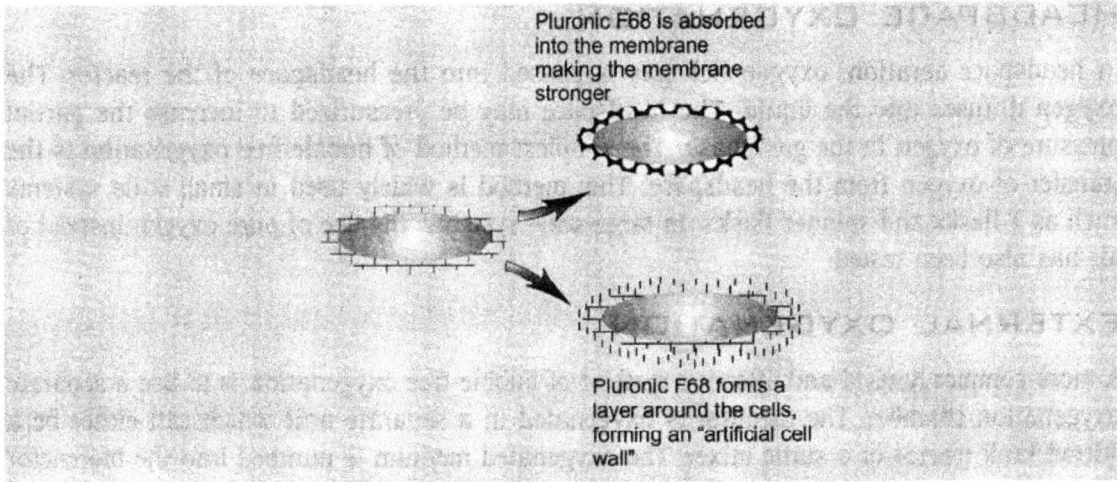

Pluronic F68 is absorbed into the membrane making the membrane stronger

Pluronic F68 forms a layer around the cells, forming an "artificial cell wall"

Figure 10.5 Shear protectant—pluronic F68

Draft tubes The important function of a draft tube is to equalize shear throughout the reactor liquid. It is likely that the insertion of a draft tube in a stirred-tank bioreactor will also have the same positive effect on animal cells.

Reducing bubble size Bubble damage is recognized as the major cause of cell damage in sparged animal cell bioreactors. When large bubbles burst, they release more energy than small bubbles. Large bubbles are therefore more destructive than small bubbles. Animal cell bioreactors are not designed to use the agitator as a tool for decreasing the bubble size diameter. The sparger therefore plays a critical role in reducing the bubble diameter. Specially designed spargers which generate very small bubbles have been designed for use in animal cell bioreactors.

Bubble-free oxygenation The importance of bubble damage as a cause of cell disruption has led to the development of bubble-free cell culture systems. There are three main techniques by which increased oxygen transfer rates can be achieved without the need for sparging.

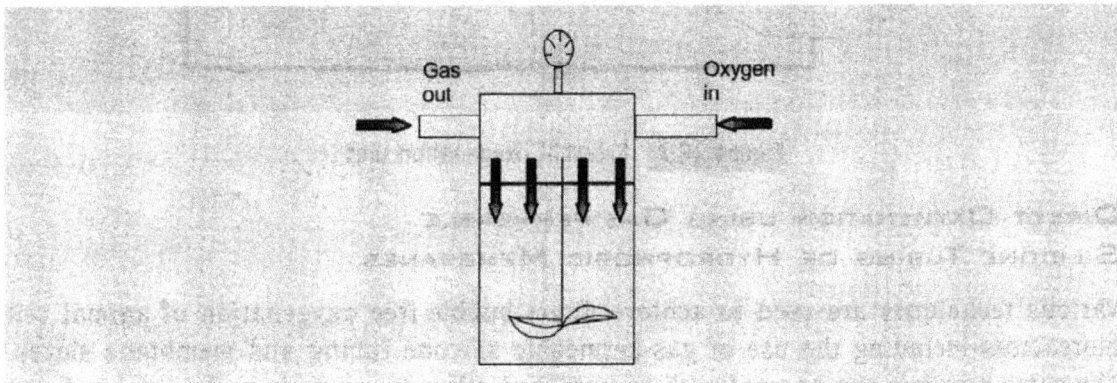

Gas out

Oxygen in

Figure 10.6 Spinner flask

HEADSPACE OXYGENATION

In headspace aeration, oxygen-rich gas is passed into the headspace of the reactor. The oxygen diffuses into the liquid. The headspace may be pressurized to increase the partial pressure of oxygen in the gas phase. The simplest method of bubble-free oxygenation is the transfer of oxygen from the headspace. This method is widely used in small-scale systems such as T-flasks and spinner flasks. In large-scale systems, the use of pure oxygen instead of air has also been tested.

EXTERNAL OXYGENATION

A more commonly used and effective method of bubble-free oxygenation is to use a separate oxygenation chamber. The medium is oxygenated in a separate unit which can either be a stirred tank reactor or a static mixer. The oxygenated medium is pumped into the bioreactor while the oxygen-depleted medium is pumped back into the oxygenation unit. A cell separation system such as a hollow fibre filter is used to separate the cells from the medium before the medium is passed into the oxygenation unit (Figure 10.7).

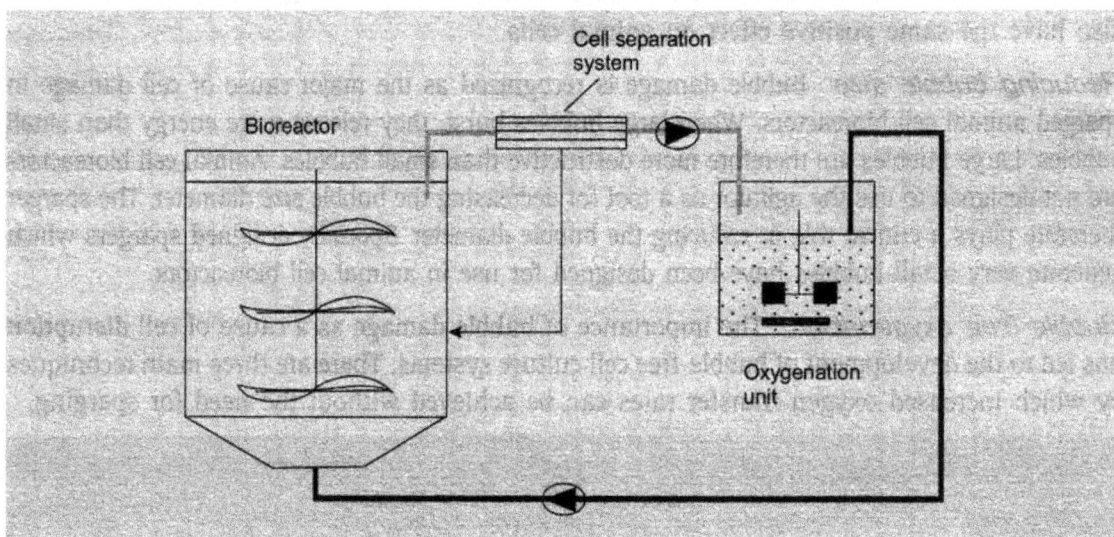

Figure 10.7 External oxygenation unit

DIRECT OXYGENATION USING GAS-PERMEABLE SILICONE TUBING OR HYDROPHOBIC MEMBRANES

Various techniques are used to achieve direct bubble-free oxygenation of animal cell bioreactors including the use of gas-permeable silicone tubing and membrane sieves. The same principle can be used with immobilized cell cultures such as fluidized bed and fibre-bed reactors (Figure 10.8). In direct oxygenation method, air is pumped through thin silicone tubing and as the gas passes along the tubing, oxygen diffuses out of the tubing and into the medium. Carbon dioxide is simultaneously transferred from the medium to the tubing. The major problem with using silicone tubing for direct oxygenation is that the diffusion rate of oxygen through the tubing is very slow and very large amounts of tubing are required for dense cell-culture systems. This in turn blocks the pores through which oxygen diffuses.

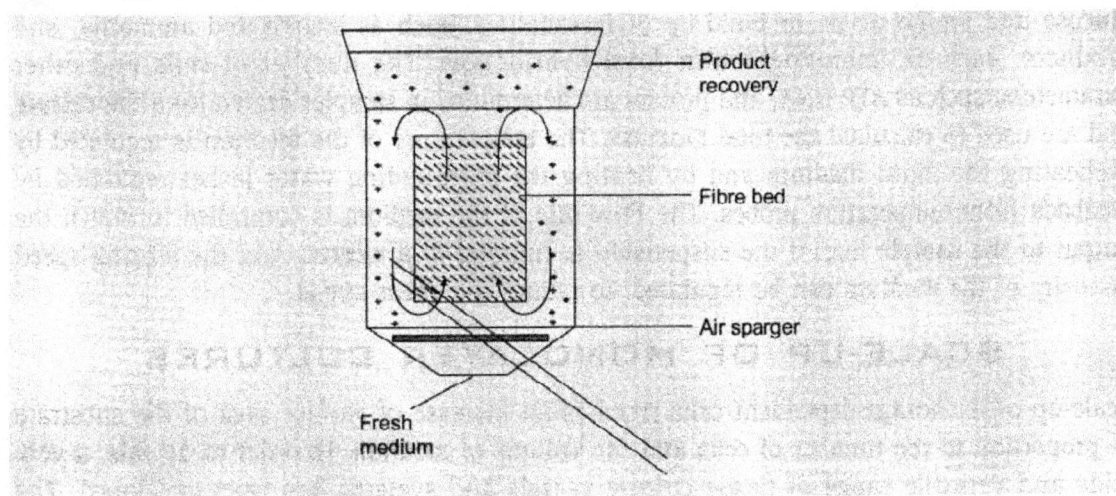

Figure 10.8 Continuous fibre bed reactor

Membrane diffusion systems operate in a manner similar to silicone tubing aeration systems. However in this case, tubes made from hydrophobic membranes are used to physically separate the gas from the culture medium. The membranes are much thinner than silicone tubing and thus offer higher oxygen diffusion rates. They are also pleated to increase the surface area for oxygen transfer. The membranes are also hydrophobic which minimizes the blockage of the membrane pores by cells.

BAG FERMENTERS

By adding platform agitation, the successful principle of plastic bioprocess containers PBCs has been developed into bag-based bioreactors. The simple, disposable nature of these units allows them easy adoption into a wide variety of facilities for a range of uses. Flat-bed bag systems exhibit low shear and reduced foaming which is normally associated with conventionally agitated systems. Bag-based fermenter technology has recently taken a further step forward by the incorporation of an impeller to offer a truly disposable stirred tank option. This greatly enhances the applicability of this format in the easy accommodation of a range of volumetric sizes in a range of institutions and facilities for a reduced capital outlay. The advent of disposable probes has also complemented the arrangement of bag systems in creating a fully disposable "plug and play" bioreactor format. This greatly facilitates the set-up in not having to sterilize or perform clean-down validation. The major disadvantage is cost.

PROCESS CONTROL

The progress of suspension culture is monitored via, pH, oxygen, CO_2, and glucose electrodes that are read from the culture *in situ*, and by assaying the utilization of nutrients, such as glucose and amino acids, or build-up of metabolites, such as lactate and ammonia, and products, such as immunoglobulin from hybridomas. The number of cells and other parameters, such as ATP, DNA, and protein are determined in samples drawn from the culture and are used to calculate the total biomass. The temperature of the medium is regulated by preheating the input medium and by heating the surrounding water jacket regulated by feedback from temperature probes. The flow rate of the medium is controlled to match the output to the sample line, if the suspension is running in a biostat. And the stirring speed viscosity of the medium can be regulated, to reduce the shear stress.

SCALE-UP OF MONOLAYER CULTURES

Scale-up of anchorage-dependent cells requires an increase of surface area of the substrate in proportion to the number of cells and the volume of medium. In order to do this, a very wide and versatile range of tissue culture vessels and systems has been developed. The methods with the maximum potential are those based on modifications to suspension culture systems because they allow a truly homogeneous unit process with enormous scale-up if time and resources allow a lengthy development period.

Advantages   Although suspension culture is the preferred method for increasing capacity, monolayer culture has the following advantages:

1. It is very easy to change the medium completely and to wash the cell sheet before adding fresh medium. This is important in many applications when the growth is carried out in one set of conditions and product generation in another. A common requirement of a medium change is the transfer of cells from serum to serum-free conditions. The efficiency of medium changing in   monolayer cultures is such that a total removal of the unwanted compound can be achieved.

2. If artificially high cell densities are needed, then these can be supported by using perfusion   techniques. It is much easier to perfuse monolayer cultures because they are immobilized and a fine filter system (to withhold cells) is not required.

3. Many cells will express a required product more efficiently when attached to a substrate.

4. The same apparatus can be used with different media/cell ratios which, of course, can be easily changed during the course of an experiment.

5. Monolayer cultures are more flexible because they can be used for all cell types. If a variety of cell types are to be used, a monolayer system might be a better investment.

6. It should be noted that the microcarrier system confers some of the advantages of a suspension culture system.

7. Monolayer cultures represent a better model for post-translational modifications of the proteins and the appropriate membrane flux, leading to the secretion of protein from the cell in a bioactive form.

Disadvantages There are four main disadvantages of monolayer compared to suspension systems:

1. They are difficult and expensive to scale-up.

2. They require much more space.

3. Cell growth cannot be monitored as effectively, because of the difficulty of sampling and counting an aliquot of cells.

4. It is more difficult to measure and control parameters such as pH and oxygen, and to achieve homogeneity.

CELL ATTACHMENT

Animal cell surfaces and the traditional glass and plastic culture surfaces are negatively charged, so for cell attachment to occur, crosslinking with glycoproteins and/or divalent cations (Ca^{2+}, Mg^{2+}) is required. The glycoprotein most studied in this respect is fibronectin, a compound of high molecular mass (220,000) synthesized by many cells and present in serum and other physiological fluids. Although cells can presumably attach by electrostatic forces alone, it has been found that the mechanism of attachment is similar, whatever the substrate charge. The important factor is the net negative charge, and surfaces such as glass and metal which have high surface energies are very suitable for cell attachment. Organic surfaces need to be wettable and negative, and this can be achieved by chemical treatment (e.g., oxidizing agents, strong acids) or physical treatment (e.g., high-voltage discharge, UV light, high-energy electron bombardment). One or more of these methods are used by manufacturers of tissue-culture-grade plastics. The result is to increase the net negative charge of the surface (for example by forming negative carboxyl groups) for electrostatic attachment. Surfaces may also be coated to make them suitable for cell attachment. A tissue-culture grade of collagen can be used for enhancing attachment.

SURFACES FOR CELL ATTACHMENT

i. Glass Alum-borosilicate glass (e.g., Pyrex) is preferred because soda-lime glass releases alkali into the medium and needs to be detoxified (by boiling in weak acid) before use.

After repeated use, glassware can become less efficient for cell attachment, but efficiency can be regained by treatment with 1 mM magnesium acetate. After several hours of soaking at room temperature, the acetate is poured away and the glassware is rinsed with distilled water and autoclaved.

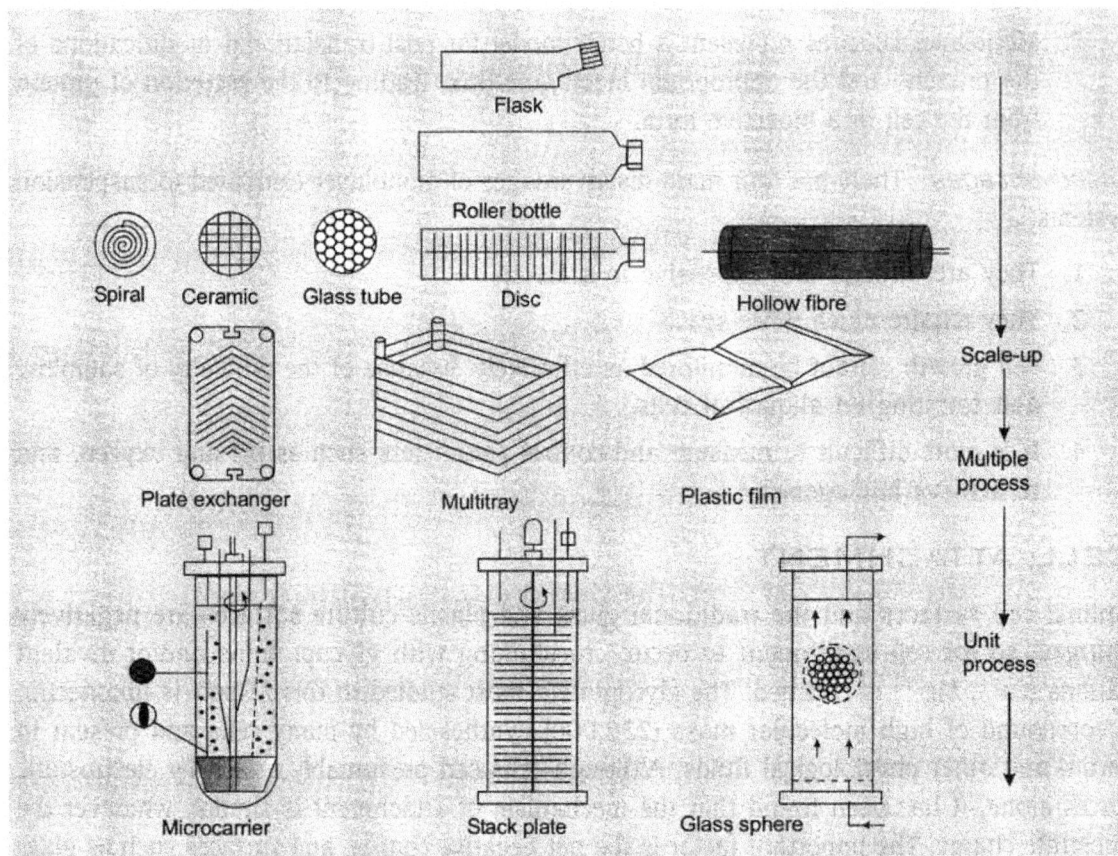

Figure 10.9 The scaling-up of culture systems for anchorage-dependent cultures (From *Culture of Animal Cells* by R.I. Freshney)

ii. Plastics   Polystyrene is the most used plastic for cell culture, but polyethylene, polycarbonate, Perspex, PVC, Teflon, cellophane, and cellulose acetate are all suitable when pre-treated correctly.

iii. Metals   Stainless steel and titanium are both suitable for cell growth because they are relatively chemically inert, but have a suitable high negative energy. There are many grades of stainless steel, and care has to be taken in choosing those which do not leak toxic metallic ions. The most common grade to use for culture applications is 316, but 321 and 304 may also be suitable. Stainless steel should be acid-washed (10% nitric acid, 3.5% hydrofluoric acid, 86.5% water) to remove surface impurities and inclusions acquired during cutting.

LARGE-CAPACITY STATIONARY CULTURES

A variety of different strategies are developed to culture adherent-dependent cells.

CELL FACTORY (MULTITRAY UNIT)

One of the simpler systems for scaling up monolayer cultures is the Nun cell factory. This system is made up of rectangular Petri-dish-like units, with a surface area of 600–24000 cm^2 interconnected at two adjacent corners by vertical tubes. This vertical aperture allows the medium to flow into all chambers only if it is positioned one over the other in a continuous way. If the flask is rotated, the medium of each compartment is isolated though allows a connection of gas phase. It can thus be thought of as a flask with a 6000 cm^2 surface area using 2 litres of medium and taking up a total volume of 12,500 ml. In practice this unit is convenient to use and produces good results, similar to plastic flasks. One of the disadvantages of the system in practice is its difficultly to wash out all the cells after harvesting with trypsin, etc. However, enough cells remain to inoculate a new culture when fresh medium is added. Given good aseptic techniques, this disposable unit can be used repeatedly. The system is used commercially for interferon production.

MULTIARRAY DISKS, SPIRALS AND TUBES

Discs, spirals, and tubes have been used to increase the surface area for monolayer growth, even for commercial use. Most matrix or multisurface propagators use the perfusion technique explained below, with an emphasis on product recovery.

ROLLER BOTTLE

The aim of scaling-up is to maximize the available surface area for growth and to minimize the volume of medium and headspace, while optimizing cell numbers and productivity. Stationary cultures have only one surface available for attachment and growth, and consequently they need a large volume of medium. The medium volume can be reduced by rocking the culture or, more usually, by rolling a cylindrical vessel. The roller bottle has nearly all its internal surface available for cell growth, although only 15–20% is covered by medium at any one time.

This system has three major advantages over static monolayer culture:

1. The increase in utilizable surface area for a given bottle.
2. The constant but gentle agitation of the medium.
3. The increased ratio of the medium's surface area to its volume, which allows enhanced gas exchange.

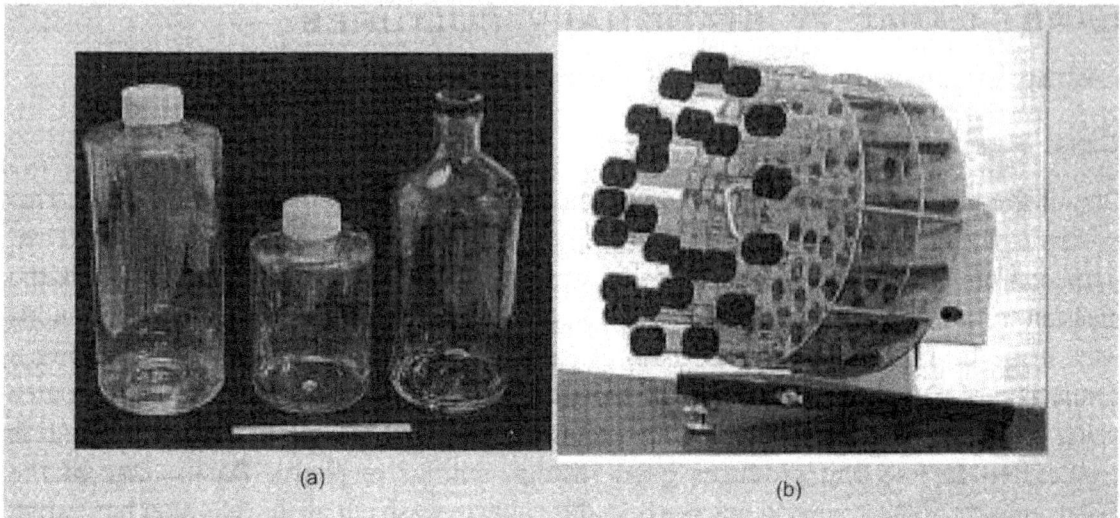

Figure 10.10 (a) Roller bottles (b) Tubes (From *Culture of Animal Cells* by R.I. Freshney)

This method reduces the volume of medium required, but still requires a considerable headspace volume to maintain adequate oxygen and pH levels. The scale-up of a roller bottle requires that the diameter is kept as small as possible. The surface area can be doubled by doubling the diameter or the length. The first option increases the volume (medium and headspace) fourfold, the second option only twofold. The only means of increasing the productivity of a roller bottle and decreasing its volume is by using a perfusion system. This is an expensive option, as an intricate revolving connection has to be made for the supply lines to pass into the bottle. However, cell yields are considerably increased and extensive multilayering takes place.

Roller bottle modifications The roller bottle system is still a multiple process, and thus inefficient in terms of staff resources and materials. To increase the surface area within the volume of a roller bottle, the following vessels have been developed.

i. SpiraCel This is available as spiral polystyrene cartridge in three sizes—3000, 4500, and 6000 cm^2. It is crucial to get an even distribution of the cell inoculum throughout the spiral, otherwise very uneven growth and low yields are achieved. Cell growth can be visualized only on the outside of the spiral, and this can be misleading if the cell distribution is uneven.

ii. Glass tubes A roller bottle is packed with a parallel cluster of small glass tubes (separated by silicon spacer rings). Three versions are available giving surface areas of 5×10^3, 1×10^4, and 1.5×10^4 cm^2. The medium is perfused through the vessel from a reservoir. The method is ingenious in that it alternately rotates the bottle 360° clockwise and then 360° anticlockwise. This avoids the use of special caps for the supply of perfused medium.

An example of its use is the production of 3.2×10^9 Vero cells ($2.3 \times 10^5/cm^2$) over six days using 6.5 litres of medium (perfused at 50 ml/min) in the 10000 cm^2 version.

iii. Increased surface area roller bottles   In place of the smooth surface in standard roller bottles, the surface is 'corrugated', thus doubling the surface area within the same bottle dimensions; e.g., extended surface area roller bottle (ESRB), or the ImmobaSil surface which is a textured silicone rubber matrix surface.

VARIOUS UNIT PROCESS SYSTEMS

There are basically three systems which fit into the fermentation apparatus (applicable for both adherent and non-adherent cells):

- cells stationary, medium moves (e.g., glass bead reactor)
- heterogeneous mixing (e.g., stack plate reactor)
- homogeneous mixing (e.g., microcarrier)

i. Bead bed reactor   This consists of a packed bed of 3–5 mm glass beads, through which medium is continually perfused. Spheres of 3-mm diameter pack sufficiently tightly to prevent the packed bed from shifting, but allow sufficient flow of medium through the column so that fast flow rates, which would cause mechanical shear damage, are not needed. The medium is transferred by a peristaltic pump in this example, but an air-lift driven system is also suitable and gives better oxygenation. The medium can be passed either up or down the column with no apparent difference in results.

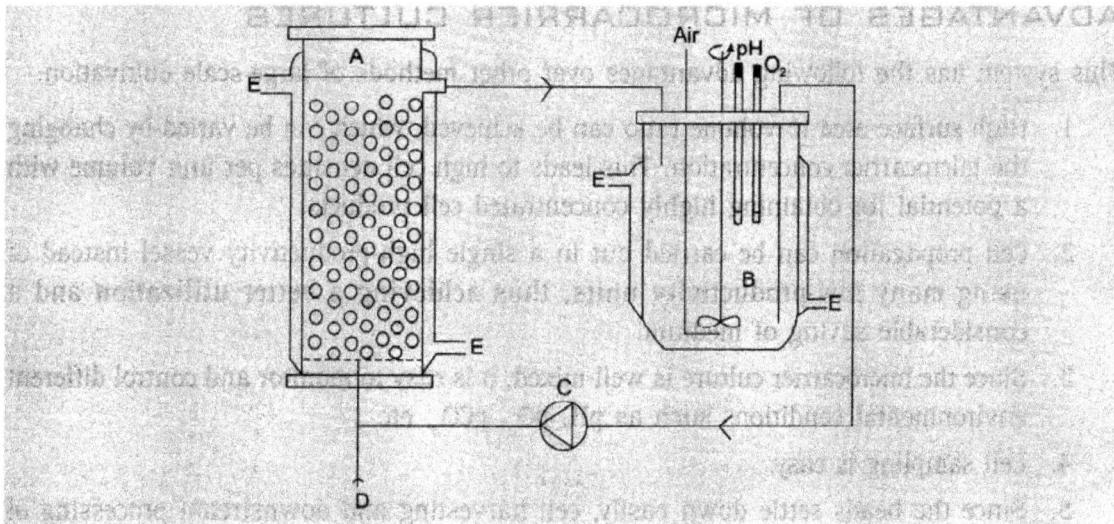

A—glass bead bed; B—reservoir; C—pump;

D—inoculation and harvest line temperature-controlled water jackets

Figure 10.11   A glass bead bioreactor (From *Culture of Animal Cells* by R.I. Freshney)

ii. Heterogeneous reactor Circular glass or stainless steel plates are fitted vertically, 5–7 mm apart, on a central shaft. This shaft may be stationary, with an airlift pump for mixing, or revolving around a vertical (6 rph.) or horizontal (50–100 rpm) axis. This multisurface propagator was used at sizes ranging from 7.5–200 litres, giving a surface area of up to 2×10^5/cm^2. The horizontal stirred plate type of vessel is easier to use and has been successful for both heteroploid and human diploid cells. The main disadvantage with this type of culture is the high ratio of medium volume to surface area (1 ml to 1–2 cm^2). This cannot be altered with the horizontal types, although it can be halved with the vertically revolving discs.

iii. Homogeneous systems (microcarrier) When cells are grown on small spherical carriers, they can be treated as a suspension culture, and advanced fermentation technology processes and apparatus can be utilized. Monolayer cultures can be grown on microbeads 90–300 µm in diameter and made of plastic, glass, collagen or dextran. The method was initiated by Van Wezel, who used dextran beads (Sephadex A-50). Culturing monolayer cells on microbeads gives a maximum ratio of the surface area of the culture to volume of the medium, up to 9000 cm^2/L based on the density and size of the beads. In this system cells can be treated as in suspension culture. The major difference from the suspension culture is in the design of the agitator, which stirs without grinding of the beads. This can be achieved with a suspended rotating pendulum or a paddle, as for suspension culture, rotating at a speed of 30 rpm.

ADVANTAGES OF MICROCARRIER CULTURES

This system has the following advantages over other methods of large-scale cultivation:

1. High surface-area-to-volume ratio can be achieved, which can be varied by changing the microcarrier concentration. This leads to high cell densities per unit volume with a potential for obtaining highly concentrated cell products.
2. Cell propagation can be carried out in a single high-productivity vessel instead of using many low-productivity units, thus achieving a better utilization and a considerable saving of medium.
3. Since the microcarrier culture is well-mixed, it is easy to monitor and control different environmental conditions such as pH, pO_2, pCO_2, etc.
4. Cell sampling is easy.
5. Since the beads settle down easily, cell harvesting and downstream processing of products is easy.
6. Microcarrier cultures can be relatively scaled up easily using conventional equipment like fermenters that have been suitably modified.

Because of the many advantages of the technique itself, it has gained great popularity. Thus, a large variety of microcarriers are available in the market.

MACROCARRIER SYSTEMS

Like microcarriers there are large porous macroscopic structures made of polylactic acid (PLA), polyglycolic acid (PGA), collagen, or gelatin (Gelfoam) in a variety of different geometries. These can be loaded with cells and stirred in a bioreactor or perfused in a fixed-bed or fluidized-bed reactor.

PERFUSED MONOLAYER CULTURE

Perfusion is frequently used to facilitate medium replacements and product recovery. The CellCube is a perfused, multisurface, single-use propagator with surface areas from 21250 to 85000 cm^2 with associated pumps, oxygenator, and system controller. Other perfusion systems use cells anchored to macrocarriers or beads in fixed-bed reactors.

i. Membrane perfusion Many systems depend on filter membrane technology in which the culture bed is a flat, permeable sheet. Membroferm is compartmentalized in such a way that the cells, medium supply, and product occupy different membrane compartments.

ii. Hollow fibre perfusion There are a number of hollow fibre perfusion systems in which adherent cells can grow on the outer surface of the perfused microcapillary bundles. High molecular weight products concentrate in the outer space with the cells, while nutrients are supplied and metabolites removed via the inner space. The potential of these systems for adherent cells lies in the re-creation of high, tissue-like cell densities, matrix interactions, and the establishment of cell polarity, all of which may be important for post-translational processing of proteins and for exocytosis. Although most suitable for adherent cells, it is extensively used for suspension cells, such as hybridomas.

iii. Opticell culture system This system consists of a cylindrical ceramic cartridge (available in surface area between 0.4–12.5 m^2) with 1 mm^2 channels running lengthwise through the unit. A medium-perfusion loop to a reservoir, (in which environmental control is carried out) completes the system. It provides a large surface area/volume ratio (40 : 1) and its suitability for virus, cell-surface antigen, and monoclonal antibody production is documented. Scale-up up to 210 m^2 is possible with multiple cartridges arranged in parallel in a single controlled unit. Cartridges are available for both attached and suspension cells, which become entrapped in the rough porous ceramic texture.

iv. Heli-Cel Twisted helical ribbons of polystyrene (3 mm × 5–10 mm ×100 µm) are used as packing material for the cultivation of anchorage-dependent cells. Medium is circulated through the bed by a pump, and the helical shape provides good hydrodynamic flow. The ribbons are transparent and therefore allow cell examination after removal from the bed.

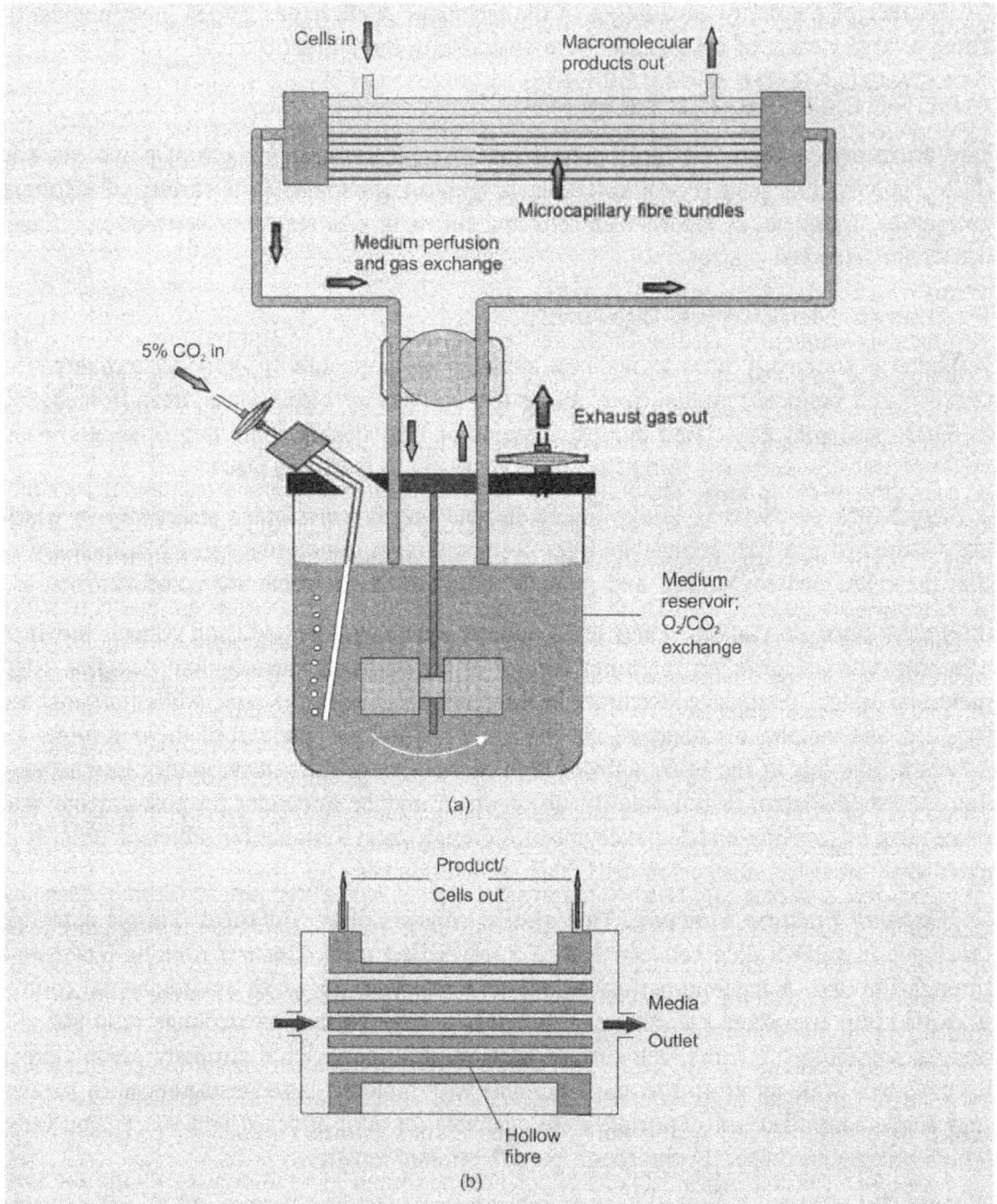

Figure 10.12 Hollow fibre perfusion unit a) showing the input and output direction b) magnified figure of the microcapillary fibre bundle mediated perfusion(From *Culture of Animal Cells* by R.I. Freshney)

FLUIDIZED-BED REACTORS

In fluidized-bed reactors, porous beads of a relatively low density—made of ceramics or a mixture of ceramics and natural products like collagen—are suspended in an upward stream of medium when the flow rate of the medium matches the sedimentation rate or circulates in a pattern as air-lift fermenter. When suspension cells lodge in the beads by entrapment, monolayer cells attach to the outer surface as well as the interstices of the porous bead or macrocarrier.

Figure 10.13 Fluidized-bed reactor (From *Culture of Animal Cells* by R.I. Freshney)

MICROENCAPSULATION

Sodium alginate behaves as a sol or gel, depending on the concentration of divalent cations. It will gel as a hollow sphere around the cells in suspension in a high concentration of divalent cations. Because the alginate acts as a barrier to high-molecular-weight molecules, macromolecules secreted by the cells are trapped within the vesicle, while nutrients, metabolites, and gas freely permeate the gel. The product and cells are recovered by reducing the concentration of divalent cations. These gels have a low immunoreactivity and can be implanted *in vivo*.

PROCESS CONTROL

Process control strategies for monolayer bioreactors are the same as that of suspension cultures. One common constraint with monolayers (especially fixed bed and hollow fibre) is the difficulty in directly monitoring the progress of culture. Cell counting and biomass determination is difficult in these types of cultures. For this, nuclear magnetic resonance (NMR) can be used to assay the contents of the culture chamber. This will help in identifying a specific metabolite. NMR spectroscope can also be used as an imaging device, producing a quasi-optical section through the chamber to reveal the distribution of the cells and for distinguishing between proliferating and non-proliferating zones of the culture.

Table 10.2   Advantages and disadvantages of various bioreactors

Bioreactor	Advantages	Disadvantages
Stirred tank	Robust design Support of large volumes Small-to large-scale production	Non linear scale-up Sterilization and clean-down validation High maintenance High capital outlay
Roller bottles	Simple Minimal control Disposable Modular Linear scale-up Small to semi-large production scale	Robotics required at large scale (capital cost) Labour-intensive at small scale Time-consuming Large footprint of roller rigs at larger scale Large incubator space required
Beg fermenter	Simple Disposable Small to semi-large production scale	Large footprint at larger scale
Multitray	Disposable Small to medium scale production Suited to cell expansion	Labour-intensive Incubator space required for larger units
Multiplate	Disposable Small to medium scale production	Difficult to assess monolayer integrity
Micro well	Versatile Limited multivariable studies	Limited throughput
Micro cassette	Versatile Rapid high throughput analysis Extensive multivariable studies	Capital cost of associated robotics

CONTINUOUS STIRRED-TANK BIOREACTOR AND ITS PROCESS CONTROL

In standard culture, known as 'batch culture', cells are inoculated into a fixed volume of medium and, as they grow, nutrients are consumed and metabolites accumulate. The environment is therefore continually changing, and this in turn enforces changes to cell metabolism, often referred to as physiological differentiation. Eventually cell multiplication ceases because of exhaustion of nutrient(s), accumulation of toxic waste products, or density-dependent limitation of growth in monolayer cultures. There are means of prolonging the life of a batch culture, and thus increasing the yield, by various substrate feed methods.

1. Gradual addition of fresh medium, thus increasing the volume of the culture (**Fed batch culture**)

2. Intermittently, by replacing a constant fraction of the culture with an equal volume of fresh medium (**semi-continuous batch**). All batch culture systems retain the accumulating waste products, to some degree, and have a fluctuating environment. All are suitable for both monolayer and suspension cells.

3. **Perfusion**, by the continuous addition of medium to the culture and the withdrawal of an equal volume of used (cell-free) medium. Perfusion can be **open**, with complete removal of medium from the system, or **closed**, with recirculation of the medium, usually via a secondary vessel, back into the culture vessel. The secondary vessel is used to "regenerate" the medium by gassing and pH correction.

4. **Continuous-flow culture**, which gives true homeostatic conditions with no fluctuations of nutrients, metabolites, or cell numbers. It depends upon the medium entering the culture with a corresponding withdrawal of medium plus cells. It is thus only suitable for suspension culture cells, or monolayer cells growing on microcarriers. Continuous-flow culture is described more fully.

CONTINUOUS-FLOW BIOREACTORS

Continuous-flow culture is the only system in which the cellular content is homogeneous, and can be kept homogeneous for long periods of time (months). This can be vital for physiological studies, but may not be the most economical method for product generation. Production economics are calculated in terms of staff time, medium, equipment, and downstream processing costs. Also taken into account are the complexity and sophistication of the equipment and process, as this affects the calibre of the staff required and the reliability of the production process. Batch culture is more expensive on staff time and culture ingredients, because for every single harvest, a sequence of inoculum build-up steps and then growth in the final vessel has to be carried out. There is also downtime as the culture is prepared for its next run. Feeding routines for batch cultures can give repeated but smaller harvests, and the longer a culture can be maintained in a productive state, the more economical the whole process becomes.

Continuous-flow culture (**stirred-tank culture**) in the **chemostat** implies that cell yields are never maximal because a limiting growth factor is used to control the growth rate.

If maximum yields are desired in this type of culture, the **turbidostat** option has to be used. Some applications, such as the production of a cytopathic virus, leave no choice other than batch culture. Maintenance of high yields, and therefore high product concentration, may be necessary to reduce downstream processing costs and these could outweigh medium expenses. For this purpose, perfusion has to be used. Although for many processes this is more economical than batch culture, it does add to the complexity of the equipment and process, and increases the risk of a mechanical or electrical failure or microbial contamination prematurely ending the production run. There is no clear-cut answer as to which type of culture process should be used—it depends upon the cell and product, the quantity of product, downstream processing problems, and product licensing regulations (batch definition of product, cell stability, and generation number). However, a relative ratio of unit costs for perfusion, continuous-flow, and batch culture in the production of monoclonal antibody is 1 : 2 : 3.5.

ADVANTAGES OF CONTINUOUS CULTURES

Continuous cultures have several advantages over batch cultures:

- In a chemostat, the cells can be maintained at a constant physiological state and growth rate. The growth rate can be adjusted by changing the feed flow rate. Consequently, it is easier to optimize productivity.

- It is not necessary to shut down the continuous fermenter as frequently as a batch fermenter. At the end of a batch fermentation, the reactor must be emptied, cleaned, sterilized and re-filled. The time required for these operations is known as turnaround time. Theoretically, a continuous fermenter could operate indefinitely without having to be shut down. In practice, however this is not possible.

- A biological phenomenon observed in batch cultures is the lag phase. The lag phase occurs at the beginning of the fermentation and represents a physiological adaptation of the culture to the new environment. Growth during this phase is very slow and the lag phase represents a period of very low productivity. Because continuous cultures are shut down with less frequency as compared to batch reactors, there is less loss of productivity during lag phases.

- Most downstream processing operations are most productive when operated in a continuous manner. Using a continuous culture allows the fermentation to be in-tune with other operations in the plant. Thus, overall plant productivity is easier to optimize.

- Continuous cultures thus offer the potential of higher productivity. As a consequence, it is possible to have smaller reactors and associated equipment and thus lower capital costs. This should lead to higher profits.

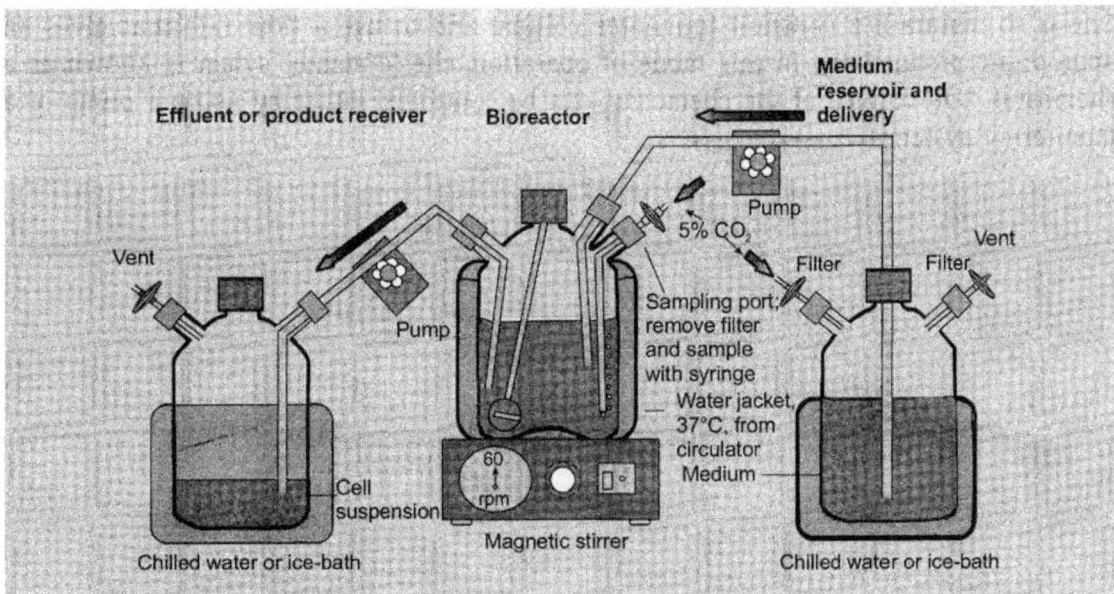

Chemostat-continuous culture (From *Culture of Animal Cells* by R.I. Freshney)

DISADVANTAGES OF CONTINUOUS CULTURES

Despite their many potential benefits, pure culture applications of continuous cultures are not well-established in industry. There are three main reasons for this:

- Risk is a major factor. The batch fermentation process is easy to understand and relatively well-established in industry. To switch from a batch to a continuous process represents a large risk; one that many managers would not take.

- The use of continuous cultures to produce many therapeutic products is not accepted by the US Food and Drug Administration as a Current Good Manufacturing Practice.

- Not all products are produced well in a continuous flow system. Many commercially important products are produced only when cell growth stops. These products are not amenable to production in a conventional continuous flow system as new biomass will not be produced to replace that removed in the effluent. Eventually the culture will be washed out. However it is possible to use non-growing cultures in immobilized cell continuous fermenters.

- Contamination of a continuous fermenter can have disastrous consequences.

MODES OF CONTINUOUS CULTURE OPERATION

An important objective of continuous culture operation is to control cell growth at a level at which productivity is optimum. There are several ways in which this can be achieved.

One is to maintain a constant fermenter volume and to use a flow rate that gives an appropriate productivity. In this mode of operation, the fermenter sytem is known as a chemostat. The volume of the chemostat can be controlled either by using a pump or a monitoring system.

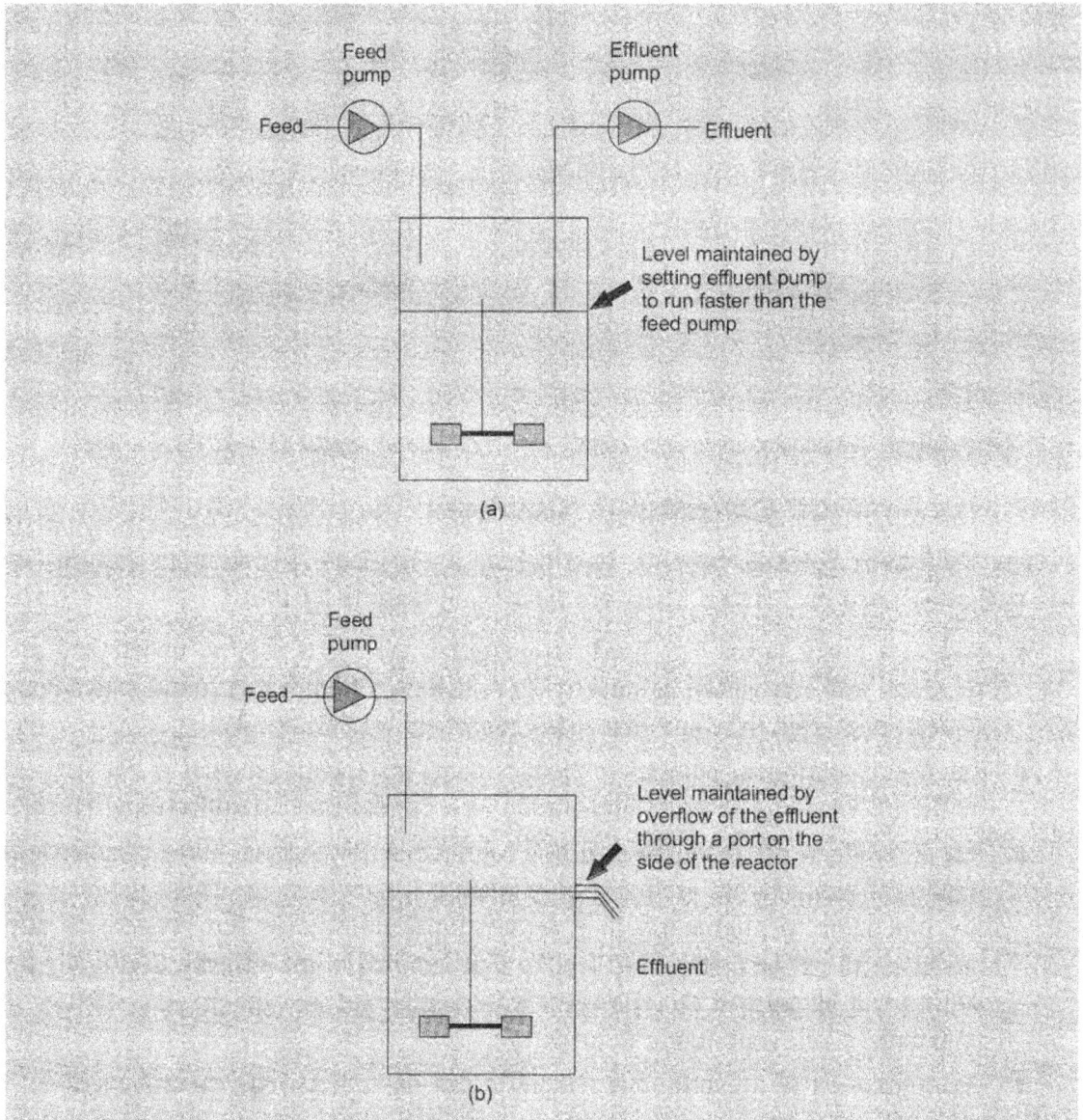

Figure 10.15 Monitoring system of a chemostat a) volume of the chemostats maintained by effluent pumping. b) volume of chemostat maintained by overflow (From *Culture of Animal Cells* by R.I. Freshney)

Chemostats The chemostat is the simplest and most common mode of operation of a continuous culture. This is a continuously stirred tank reactor with an inlet and outlet so that medium can be passed continuously at a fixed rate keeping the volume constant. The theory for chemostats was established by Monod, Novick and Szillard independently in 1942. The culture is first grown in batch mode and when cells reach the log phase they are grown in continuous mode. The medium employed has one growth-limiting nutrient and the rest are in excess. Hence cell growth is proportional to the growth-limiting nutrient. Steady state is obtained by a continuous and constant feed and harvesting rate where the dilution rate is determined by growth rate.

Where contamination can be a significant problem, a pump-based control system is preferred. This set-up is commonly used in laboratory investigations and animal cell culture systems. An overflow system has the advantage in that only one pump is required. However as the effluent flow rate is determined by gravity alone, there is a greater possibility of contaminants moving up the effluent tube into the reactor. Overflow systems are however widely used in wastewater treatment and have been used in the large-scale continuous culture of bacteria.

In other techniques, a fermenter variable, e.g., turbidity or pH, will be monitored using an appropriate detector and the liquid flow rate will be automatically adjusted so as to maintain the variable at a constant level. Examples of these types of continuous fermenters are the pH-stat, **turbidostat** and nutristat. Apart from the pH-stat, these reactors are however rarely used as the necessary measurement-control systems since they are generally unreliable over long periods of time.

PROCESS CONTROL

The progress of suspension culture is monitored via pH, oxygen, CO_2, and glucose electrodes that read from the culture *in situ*, and by assaying the utilization of nutrients, such as glucose and amino acids, or build-up of metabolites such as lactate and ammonia, and products such as immunoglobulin from hybridomas. The number of cells and other parameters, such as ATP, DNA, and protein are determined in samples drawn from the culture and are used to calculate the total biomass. The temperature of the medium is regulated by preheating the input medium and by heating the surrounding water jacket regulated by feedback from temperature probes. The flow rate of the medium is controlled and is matched to the output to the sample line, if the suspension is running in a biostat. And the stirring speed can be regulated, along with the viscosity of the medium, to reduce the shear stress.

Process control strategies for monolayer bioreactors are same as that of suspension cultures. One common constraint in monolayer cultures (especially fixed bed and hollow fibre) is the difficulty in directly monitoring the progress of culture. Cell counting and biomass determination are difficult in these types of cultures. For this, nuclear magnetic resonance (NMR) can be used to assay the contents of the culture chamber. This will help in identifying a specific metabolite. NMR spectroscope can also be used as an imaging device, producing a quasi-optical section through the chamber to reveal the distribution of the cells and even for distinguishing between proliferating and non-proliferating zones of the culture.

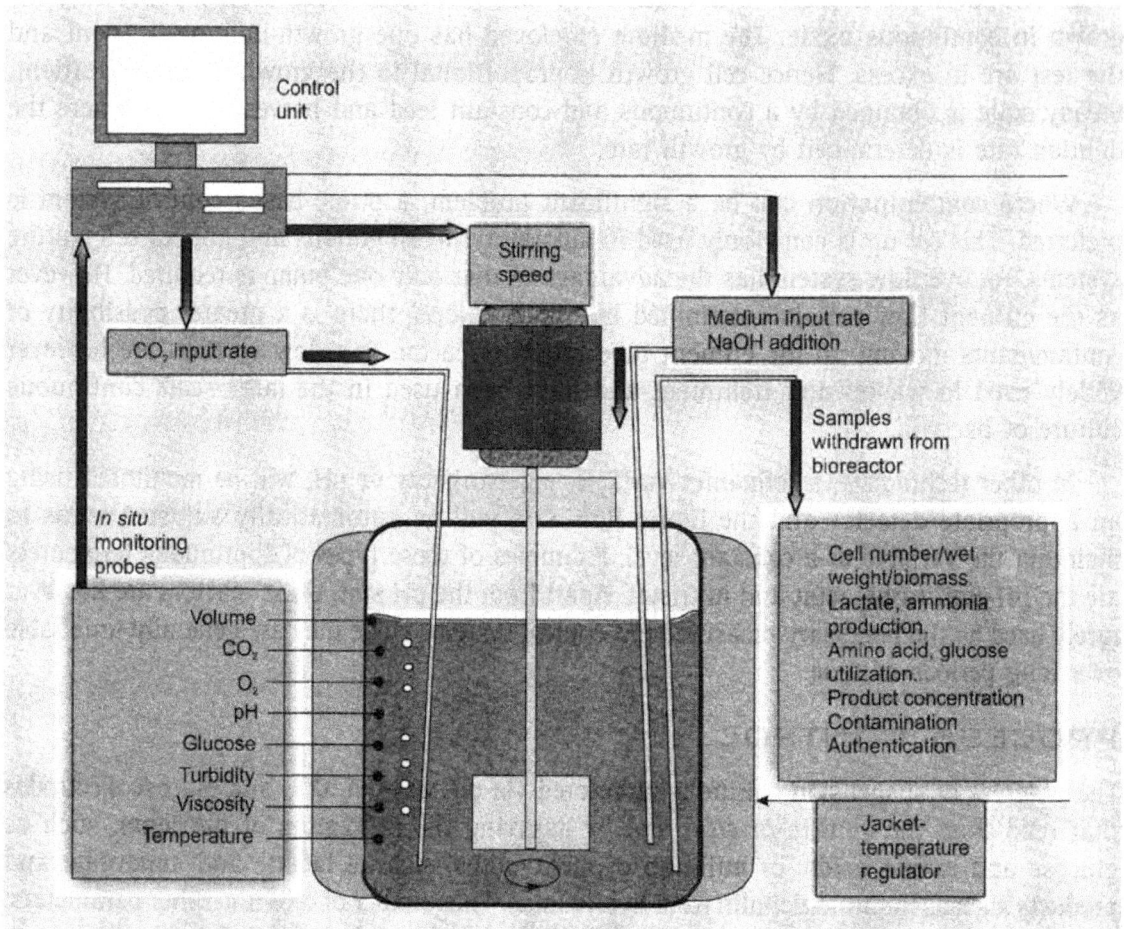

Figure 10.16   Bioreactor-prototype (From *Culture of Animal Cells* by R.I. Freshney)

AIR-LIFT BIOREACTOR

Air-lift bioreactor is one in which circulation of the culture medium and aeration is achieved by injection of air into some lower part of the fermenter. Air-lift fermenters have a cylinder inside the culture vessel through which air is passed for circulating the medium. It is a very well-aerated system. Though usually it is not suitable for animal cell production due to shear damage to delicate animal cells, it is used mainly for monoclonal antibody production.

There are two main designs—bubble column fermenter and air-lift fermenter.

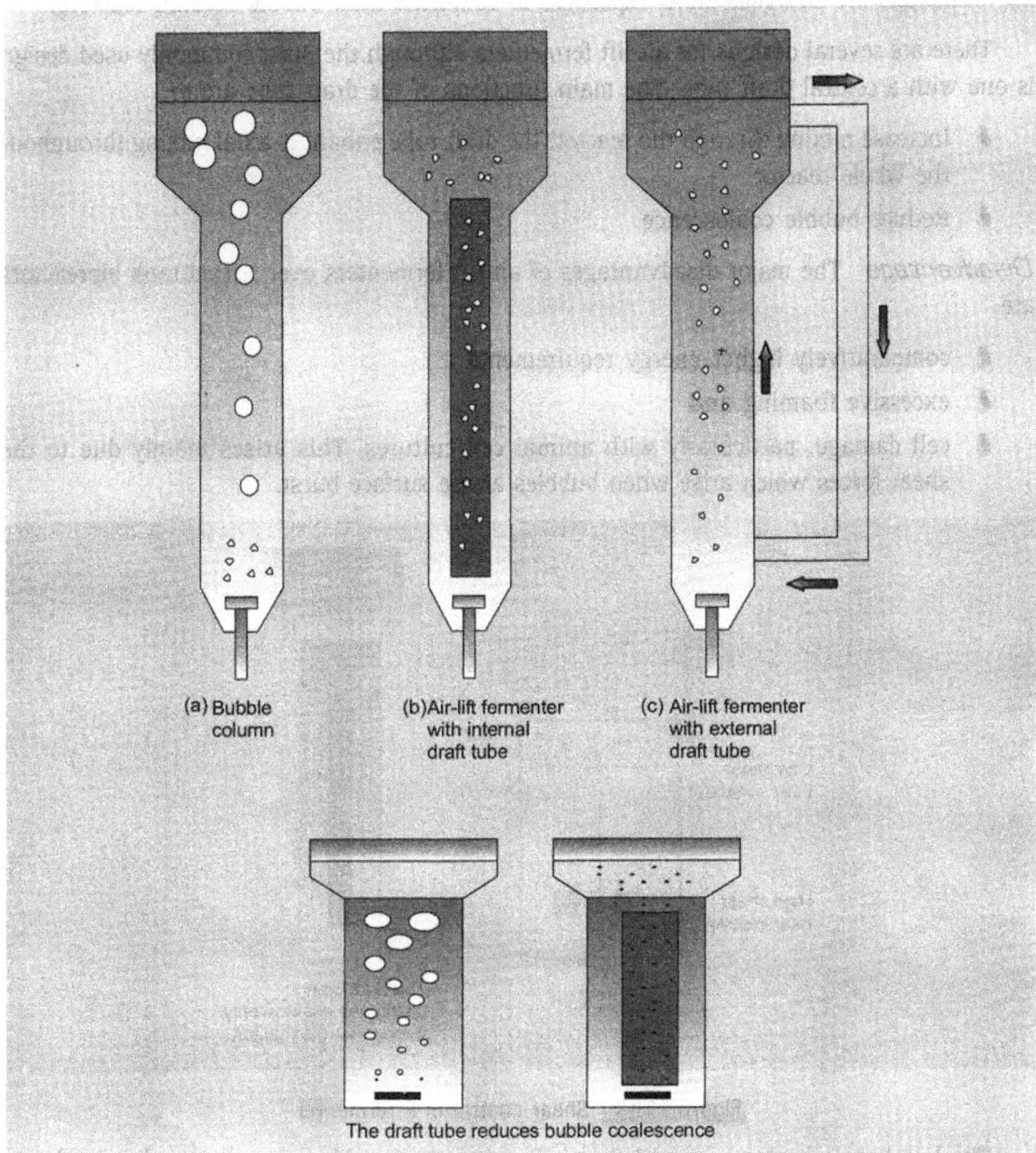

(a) Bubble column

(b) Air-lift fermenter with internal draft tube

(c) Air-lift fermenter with external draft tube

The draft tube reduces bubble coalescence

Figure 10.17 Fermenter types

Sparging without mechanical agitation can also be used for aeration and agitation. Bubble-driven bioreactors are commonly used in the culture of shear-sensitive cells. An air-lift fermenter differs from bubble column bioreactors by the presence of a draft tube which provides better mass and heat transfer efficiencies and more uniform shear conditions. Air-lift fermenters are however more expensive to construct than bubble column reactors.

There are several designs for air-lift fermenters although the most commonly used design is one with a central draft tube. The main functions of the draft tube are to:

- Increase mixing through the reactor; the draft tube enhances axial mixing throughout the whole reactor
- Reduce bubble coalescence

Disadvantage   The major disadvantages of air-lift fermenters over stirred tank bioreactors are:

- comparatively higher energy requirements
- excessive foaming and
- cell damage, particularly with animal cell cultures. This arises mainly due to the shear forces which arise when bubbles at the surface burst.

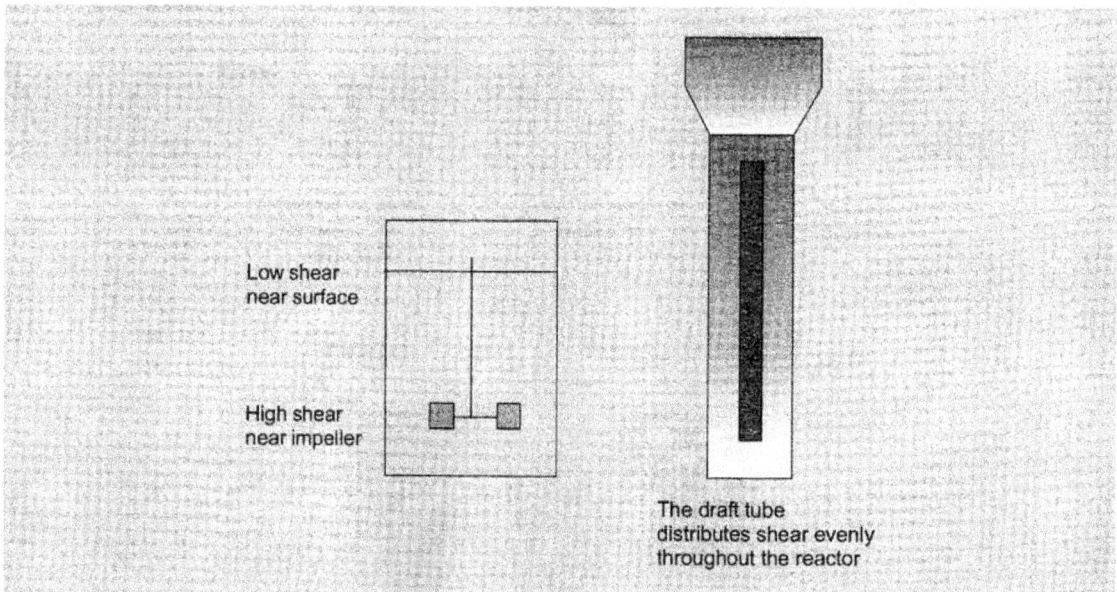

Low shear near surface

High shear near impeller

The draft tube distributes shear evenly throughout the reactor

Figure 10.18   Shear control in a fermenter

The low bubble coalescence and thus cell stress presumably occur due to the circulatory effect that the draft tube induces in the reactor. Circulation occurs in one direction and hence the bubbles also travel in one direction. Small bubbles lead to an increased surface area for oxygen transfer and equalize shear forces throughout the reactor. This is believed to be the major reason why air-lift bioreactors have higher productivities than stirred-tank reactors.

An air-lift reactor is divided into three regions:

1. the air-riser,
2. downcomer and
3. the disengagement zone.

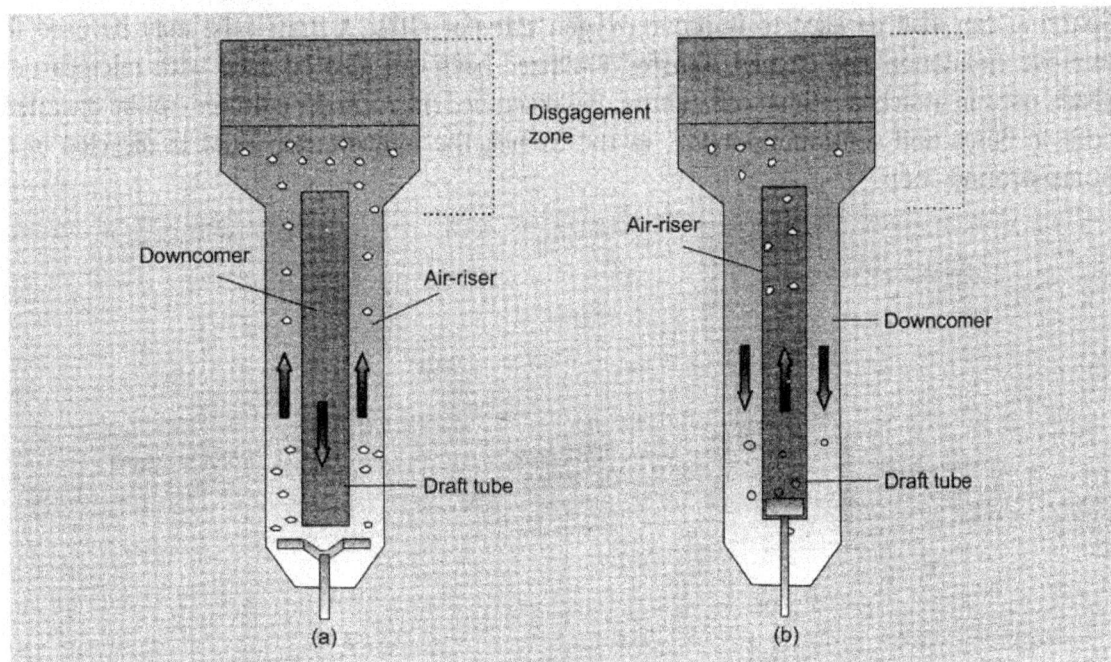

Figure 10.19   Air-lift fermenter a) small-scale fermenter b) large-scale fermenter.

The region into which bubbles are sparged is called the air-riser. The air-riser may be on the inside or the outside of the draft tube. The latter design is preferred for large-scale fermenters as it provides better heat transfer efficiencies. The rising bubbles in the air-riser cause the liquid to flow in a vertical direction. To counteract these upward forces, liquid will flow in a downward direction in the downcomer. This leads to liquid circulation and thus improved mixing efficiencies as compared to bubble columns. The enhanced liquid circulation also causes bubbles to move in a uniform direction at a relatively uniform velocity. This bubble flow pattern reduces bubble coalescence.

The role of the disengagement zone is to add volume to the reactor, reduce foaming and minimize recirculation of bubbles through the downcomer.

FLUIDIZED-BED REACTOR

Fluidized-bed bioreactors maintain high biomass concentrations and at the same time good mass transfer rates in continuous cultures. Fluidized-bed bioreactors are an example of reactors in which mixing is assisted by the action of a pump. In a fluidized-bed reactor, cells or enzymes are immobilized in and/or on the surface of light particles. A pump located at the base of the tank causes the immobilized catalysts to move with the fluid. The pump pushes the fluid and the particles in a vertical direction. The upward force of the pump is balanced by the downward movement of the particles due to gravity. This results in good circulation. Sparging can also be used to improve oxygen transfer rates. A draft tube may be used to improve circulation and oxygen transfer. Fluidized beds can also be used with microcarrier beads used in attached animal cell culture. Fluidized-bed microcarrier cultures can be operated both in batch and continuous mode. In the former, the fermentation fluid is recycled in a pump-around loop.

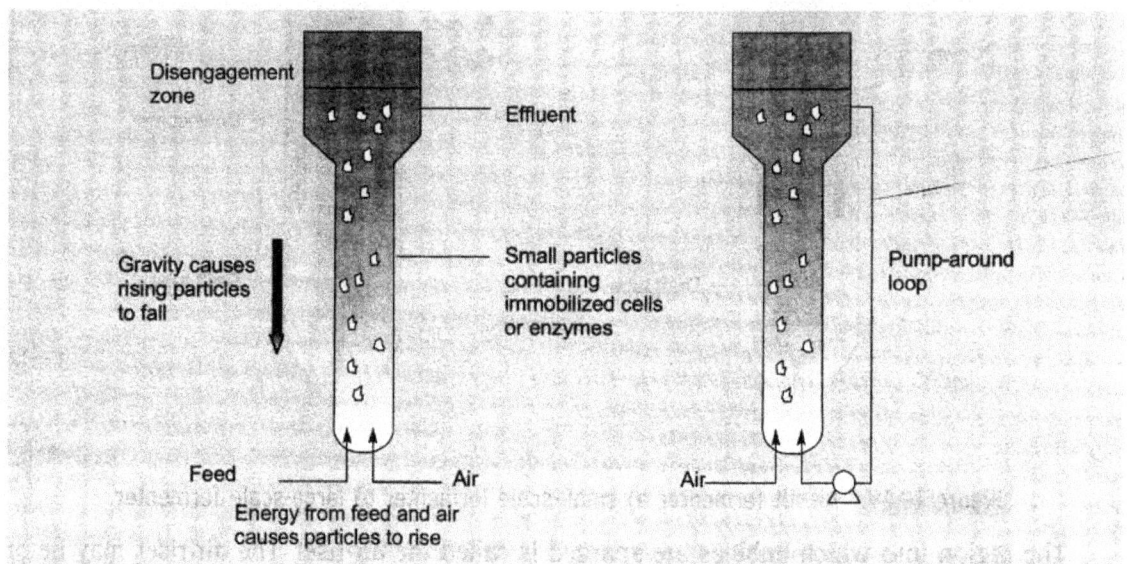

Figure 10.20 Fluidized-bed reactor a) Gravity-driven recycling of fermentation fluid b) Pump around loop-mediated recycling of fermentation fluid.

ENCAPSULATION AND PACKED-BED REACTORS

One of the methods for minimizing cell damage is immobilization. Cells can be immobilized onto microcarrier beads or in hollow fibre systems. However, not all cells or cell culture processes are amenable to immobilization. One of the common cell immobilization methods is encapsulation.

Micro-encapsulation is a process in which cells are surrounded by a coating to give small capsules with many useful properties. In its simplest form, a **microcapsule** is a small sphere with a uniform wall around it. The material inside the microcapsule is referred to as the core, internal phase, or fill, whereas the wall is sometimes called a shell, coating, or membrane. Most microcapsules have diameters between a few micrometres and a few millimetres.

Reasons for Encapsulation

The reasons for microencapsulation are countless. In some cases, the core must be isolated from its surroundings to isolate sensitive cells and protect it from the deteriorating effects of the culture conditions, retarding evaporation of a volatile core, for immobilization of cells, for concentrating cell for a high density, etc.

Many of the methods of encapsulation damage mammalian cells. Some common methods for encapsulation are:

Interfacial polymerization In interfacial polymerization, the two reactants in a polycondensation meet at an interface and react rapidly. The basis of this method is the classical Schotten–Baumann reaction between an acid chloride and a compound containing an active hydrogen atom, such as an amine or alcohol, polyesters, polyurea or polyurethane. Under the right conditions, thin flexible walls form rapidly at the interface. A solution of the pesticide and a diacid chloride are emulsified in water and an aqueous solution containing an amine and a polyfunctional isocyanate is added. A base is present to neutralize the acid formed during the reaction. Condensed polymer walls form instantaneously at the interface of the emulsion droplets.

In situ polymerization In a few microencapsulation processes, the direct polymerization of a single monomer is carried out on the particle surface. In one process, for example, cellulose fibres are encapsulated in polyethylene while immersed in dry toluene. Usual deposition rates are about 0.5 µm/min. Coating thickness is in the range of 0.2–75 µm. The coating is uniform, even over sharp projections.

Matrix polymerization In a number of processes, a core material is imbedded in a polymeric matrix during formation of the particles. A simple method of this type is spray-drying, in which the particle is formed by evaporation of the solvent from the matrix material. However, the solidification of the matrix can also be caused by a chemical change.

Various designs of bioreactors are available with entrapped cultures:

Opticell The special ceramic cartridges for suspension cells (S Core) entrap the cells within the porous ceramic walls of the unit. They are available in sizes from 0.42 m^2 to 210 m^2 (multiple cartridges), which will support 5×10^{10} cells with a feed/harvest rate of 500 litres/day and give a yield of about 50 g of monoclonal antibody per day.

Fibres A simple laboratory method is to enmesh cells in cellulose fibres (DEAE, TLC, QAE. TEAE). The fibres are autoclaved at 30 mg/ml in PBS, washed twice in sterile PBS, and added to the medium at a final concentration of 3 mg/ml in a spinner-stirred bioreactor. This method has even been found suitable for human diploid cells.

Porous carriers Microcarriers and glass spheres are restricted to attached cells and, because a sphere has a low surface-area-to-volume ratio, they are restricted in their cell density potential. Their advantages are summarized below. A change from a solid to porous sphere of open, interconnecting pores increases their potential enormously. There are various types of porous (micro) carriers commercially available. A characteristic of these porous carriers is their equal suitability for suspension cells (by entrapment) and anchorage-dependent cells (huge surface area). The problem with many immobilization materials is that diffusion paths become too long, preventing scale-up. A sphere is ideal in that cells and nutrients have to only penetrate 30% of the diameter to occupy 70% of the total volume. This facilitates scale-up as each sphere, whether in a stirred, fluidized, or fixed-bed culture, can be considered an individual mini-bioreactor.

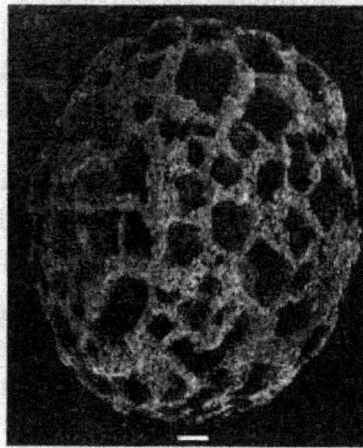

Figure 10.21 Porous microcarrier glass sphere

Advantages of porous carriers compared to solid carriers

1. Unit cell density 20- to 50-fold higher.
2. Support both attached and suspension cells.
3. Immobilization in 3D configuration easily achieved.
4. Short diffusion paths into a sphere.
5. Suitable for stirred, fluidized, or fixed-bed reactors.
6. Good scale-up potential by comparison with analogous systems (e.g., microcarrier at 4000 litre).
7. Cells protected from shear.
8. Capable of long-term continuous culture.

FIXED-BED (POROSPHERE)REACTOR

The fixed bed can be prepared with solid spheres and porous Siran spheres (Schott) of 4–6 mm diameter. The only differences are that the bed should be packed with oven-dried spheres and the void volume of cells plus medium inoculated (2×10^6/ml) directly to the bed (dry beads permit better penetration of cells). A larger medium volume is required, and faster perfusion rates (5–20 linear cm/min) should be used (cells are protected from medium shear within the matrix), e.g., 1-litre packed bed needs at least a 15-litre reservoir. After the initial 72 h-period, 10 litres of fresh medium is added daily.

Typical results are 2.75×10^{10} viable cells/litre giving an average yield of 166 mg/litre/ day (compared to stirred reactor and air-lift cultures of the same hybridoma of 25.5 and 18.5 mg/litre respectively). This method is a low-investment introduction to high productivity production of mAb which is simple to use and reliable with low maintenance, at least for the first 50 days of culture.

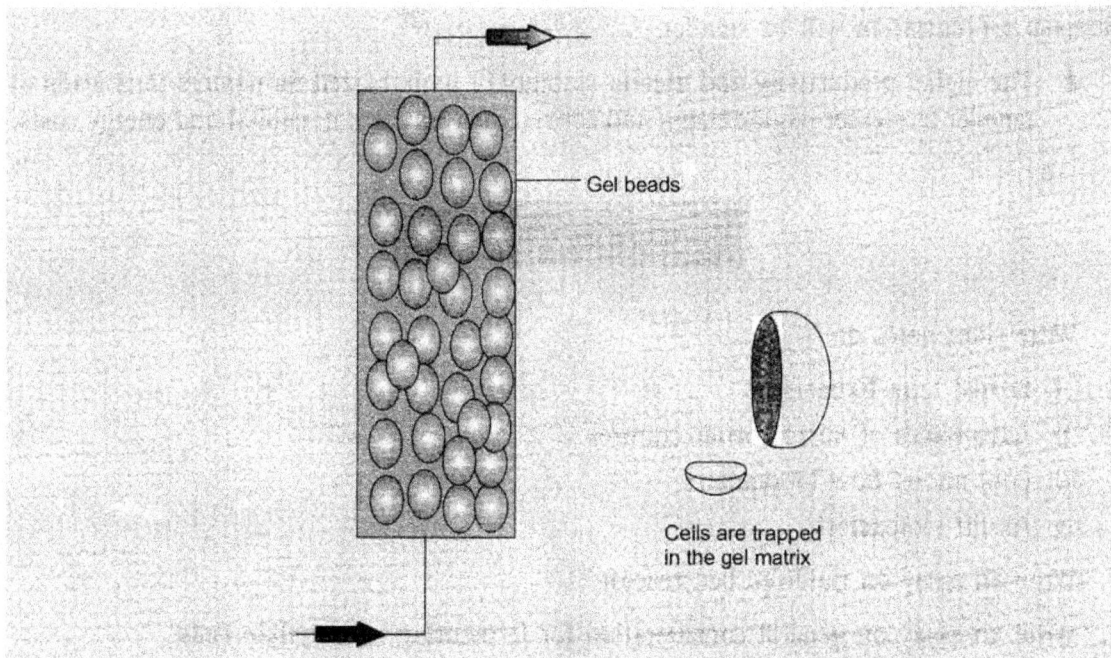

Figure 10.22   Fixed-bed reactor showing gel matrix and cell trapment

STIRRED CULTURES

The Cultispher-G (gelatin), Cellsnow (cellulose), and ImmobaSil (silicone rubber) microcarriers are the most suited to stirred bioreactors, and can be used in an identical manner to solid microcarriers, i.e., 2 g/litre in shake flasks, spinner flasks, or stirred fermenters. The silicone rubber of the ImmobaSil microcarriers facilitates oxygen diffusion and this offers a great advantage over other formulations. About a 40-fold higher concentration of attached cells, and even greater densities of suspension cells, can be achieved over solid microcarriers. The cells can be released from the microcarriers by collagenase treatment.

Advantages of immobilized cell reactors The ability of maintaining high cell concentrations in the reactor at high dilution rates provides immobilized cell reactors a number of advantages over chemostats.

- More cells means that the fermenter contains more catalysts and therefore high conversion rates can be achieved.
- Immobilized cell fermenters are also more stable than chemostats.

In a chemostat, a temporary (transient) increase in the dilution rate will cause a rapid drop in cell number. The entry of a slug of toxic substances in the feed will have the same effect. It will take time for the cell number to build up again. Since the cells are not as easily washed out of an immobilized cell reactor, the recovery time will be quicker and the fall in biomass concentration will be smaller.

- The higher productivity and greater stability of immobilized fermenters thus leads to smaller fermenter requirements and considerable savings in capital and energy costs.

REVIEW QUESTIONS

1. Write short notes on:
    i. Stirred tank bioreactors
    ii. Advantages of micro carrier cultures
    iii. Continuous flow bioreactors
    iv. Air-lift bioreactors

2. Write an essay on fluidized bed reactor.

3. Write an essay on general consideration for fermenting mammalian cells.

4. What are the methods used for large capacity stationary cultures?

11

GENERAL STRATEGY FOR THE PRODUCTION OF MONOCLONAL ANTIBODY

Serum contains a mixture of antibodies recognizing many different antigens. Furthermore, different epitopes of any particular antigen are also recognized. These non-specific reactions and cross-reactivity may limit the usefulness of antisera. This heterogeneity is the result of antibodies being the products of clonal lineages of mature B cells. Each clonal lineage of a B cell produces a specific antibody that recognizes a specific epitope. It is not possible to isolate and grow B cells in culture. In 1975 Kohler and Milstein described a technique that allowed the growth of clonal populations of cells secreting antibody of defined specificity, for which they received the Nobel prize in medicine in 1984. The method consists of fusing B-cells with myeloma cells (a type of B-cell tumour). The fused cell is known as a **hybridoma**. This hybridoma now has the ability to grow in culture and it secretes an antibody of defined specificity (phenotype acquired from the B cell). The antibody produced by the hybridoma is referred to as a monoclonal antibody (mAb).

The major advantage of a mAb is its unlimited supply. The hybridoma cell line has indefinite life whereas only a limited amount of serum can be obtained from immunized animals. In addition, the mAb is a defined reagent that recognizes a single epitope. Individual animals may respond differently to the same antigen resulting in potentially different polyclonal antibodies against the same antigen. Another advantage is the ability to produce specific antibody with impure antigen. A disadvantage of mAbs is that only a single epitope is recognized and the mAb may not have the desired specificity or affinity. Secondly, mAbs are time-consuming (6–9 months) to produce and require tissue culture facilities. Following immunization, there are several steps involved in the production of monoclonal antibodies.

1. Immunization and preparation of spleen from a suitable animal
2. Fusion
3. Selection of hybrid / fused cells
4. Screening

5. Cloning

6. Production of monoclonal immunoglobulins

Box 11.1   Characterization of monoclonal immunoglobulins

- Foreigness
- Molecular complexity
- Molecular size
- B-cell epitope
- T-cell epitope
- Class II binding site
- Degradation/presentation
- Particulate/phagocytosis

IMMUNIZATION

Immunized animals are needed for the first step in mAb production. There is no set rule for selecting a particular immunization schedule as it may vary for a given antigen.

IMMUNOGEN PREPARATION

The preparation of the immunogen and the immunization protocol can affect the nature of the antibody produced. Typically a purified protein is used as an immunogen. Ideally an immunogen (Table 11.1) should be purified to homogeneity for the production of monospecific antisera. Partially purified or crude immunogens can also cause problems with suppression and 'antigenic competition' in that some proteins are more immunogenic than others. In some situations it is not necessary to use highly purified immunogens. For example, it is also possible to use animals immune to an infectious disease or to immunize with crude antigens. Monoclonal antibodies with the desired specificities are then selected during the screening stage.

Synthetic peptides and recombinant proteins can be used in situations where it is not possible to purify sufficient protein of the desired purity. In addition, the use of recombinant proteins or synthetic peptides will permit the production of antibodies against particular epitopes or domains of the protein. Synthetic peptides or other small chemicals, often called **haptens**, need to be coupled to a carrier, or larger protein, before immunization. Conjugation to a carrier ensures an adequate immunogen size, increased haptenic epitope density, and source of heterologous T-cell epitopes. The haptens or peptides are chemically cross-linked to the carrier. The recombinant DNA approach (i.e., genetic engineering) is also amenable to the inclusion of good T-cell epitopes and class II MHC binding sites.

Native proteins tend to be more immunogenic than denatured proteins, because native structure of antigen epitope tends to produce antibody against the conformation epitope.

Immunization with denatured proteins tends to produce antibodies against sequential or linear determinants. But it is generally easier to obtain a higher degree of purity under denaturing conditions.

Table 11.1 Different animals and birds for production of antibodies

Animal	Advantages	Disadvantages
Rabbits	Large amounts of sera	Require more immunogen
	Strong antibody response	High background
	Easy to maintain	
Mice/Rats	Inbred strains	Small amounts of sera
	Monoclonal antibodies possible	
	Easy to maintain	
Guinea pigs	Generally high titer antibodies	Difficult to bleed
	Easy to maintain	Small amount of sera
Goats/Sheep	Very large amounts of sera	Difficult to maintain
Chickens	Good for some highly conserved mammalian antigens	Difficult to maintain

CHOICE OF ANIMAL

A wide range of vertebrate species can be used for the production of antibodies, but the two most often used animals are mice and rabbits. Rabbits react to most immunogens and relatively large quantities of antisera are obtained. The main disadvantage is that larger quantities of antigen are needed and rabbits often exhibit a high background response to heterologous antigens.

Rodents, especially mice, are almost always the animals of choice. Mice are easy to handle and several different genetic strains are available and hence are widely used in hybridoma technology. This is primarily due to the availability of mouse myeloma cell lines. Human and simian hybridomas have been successfully produced by transforming the lymphocytes with Epstein–Barr virus. In hybridoma technology, in future, if we have plans to produce ascitic tumours, the mouse strain used for the immunization should match the mouse strain from which the myeloma cells were derived. For stable hybrid clones BALB/c mice are preferred for immunization because all the mouse myeloma cell lines are derived from this strain. Moreover hybrid cell clones obtained with BALB/c lymphocytes will grow as tumours in this strain when injected intraperitoneally, thus generating high-titred ascitic fluid.

The different genetic strains of mice provide a more uniform response to immunogens and also some control of the immune response since the different strains may respond differently to the same antigen. In addition, mice are used as models for many infectious diseases. Adult animals should be used for immunizations (i.e., rabbits 6–10 lb and mice > 7 weeks). If possible, 2–3 rabbits and 5–10 mice should be immunized with the same immunogen to control animal-to-animal variation.

ADJUVANTS

Adjuvants are substances that enhance the overall immune response. Most adjuvants contain two components. One component will form a deposit that prevents the rapid breakdown of the antigen and promotes a slow release of the antigen, called the **depot effect**. There are several different classes of adjuvants with distinct mechanisms by which the depot effect is evoked.

The second component of many adjuvants non-specifically stimulates the immune system (mitogenic or polyclonal activation). For example, killed bacteria, especially mycobacteria, or bacterial components are included in some adjuvants. Common bacterial components included in adjuvants are lipopolysaccharide (LPS) or muramyldi-dipeptide (MDP), a synthetic analogue of the adjuvant-active component from the cell walls of mycobacteria. These substances stimulate macrophages and other immune effector cells to release cytokines which then enhance the antibody response. The immunomodulatory component may also influence the class of immunoglobulin produced as well as the isotype of IgG.

Box 11.2	Merits of different routes of immunizations	
Route		**Comments**
Subcutaneous	sc	Easy injections
Intramuscular	im	Slow release
Intradermal	id	Difficult injections
Intravenous	iv	Slow release
		Only for boosting
Intraperitoneal	ip	No adjuvants
		Easy injections
		Common in mice

INJECTION

It is possible to immunize by several different routes. The route of injection can also affect the antibody response, both in terms of quantity of antibody produced and in the qualitative aspects of the antibody. The volume of immunogen and how quickly the immunogen should be released into the lymphatics or circulation will influence the choice of injection method. Rabbits are generally immunized subcutaneously at multiple sites to stimulate regional lymph nodes. Immunization of mice with large volumes can only be accomplished intraperitoneally. Intramuscular and intradermal injections are best for slow release. Intravenous immunization is only used for boosting, and immunogens containing adjuvants should not be given intravenously.

Dose

The antigen dose will depend on the animal and on the purity and the immunogenicity of the antigen. In general, 0.05–1 mg and 5–50 µg of pure antigen are needed for rabbits and mice, respectively. To obtain the optimal antibody response (i.e., highest titer and affinity) the animals will need to be re-immunized, or boosted. Boosting promotes class switching (e.g., IgM → IgG), higher levels of antibody, and an increase in the affinity of the antibody for the antigen. Typically, boosting is carried out at intervals of 2–6 weeks, and 3–4 total immunizations are given.

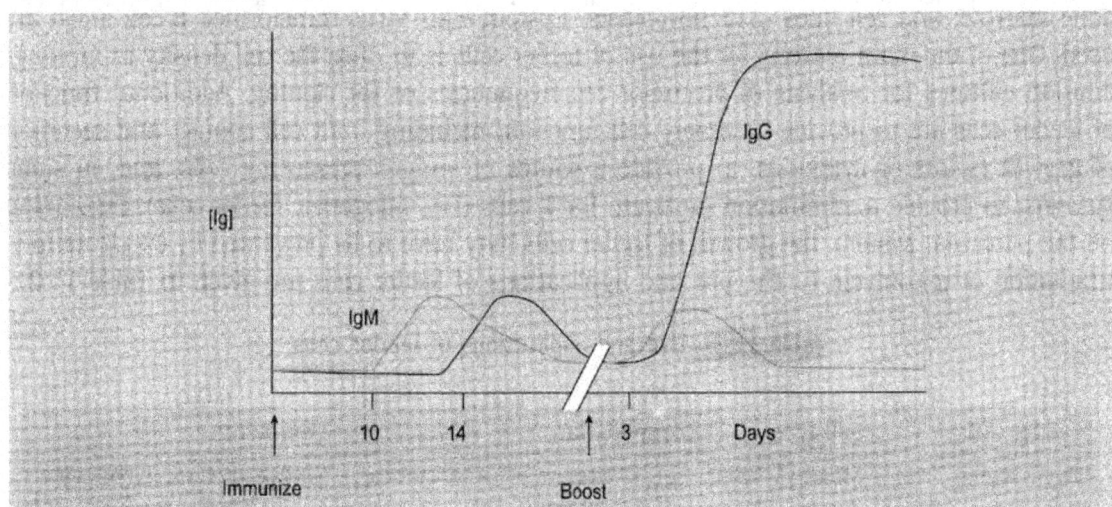

Figure 11.1 Typical immune response following immunization

The mice are immunized, boosted and tested until an antibody response of the desired specificity is elicited. Generally the monoclonal antibodies obtained will be reflective of the antibodies found in the polyclonal sera. Therefore before proceeding with the fusion, it is crucial to ensure that the mouse exhibits antibodies of the desired specificity.

Preparation of Spleen

A selected immunized mouse is killed by ether anaesthesia, the spleen is removed aseptically and placed in a Petri dish having 10 ml of RPMI-1640 medium. A cell suspension is made by teasing the spleen with a pair of curved forceps, clumps and membrane fragments allowed to settle and the resulting cells are diluted/adjusted to 1×10^7 using RPMI-1640 medium. One healthy spleen can yield on an average $100–200 \times 10^6$ cells.

MYELOMAS

Myelomas are induced in BALB/c mouse by injecting intraperitoneally either mineral oil or pristane (2,6,10,14-tetramethylpentadecane) at 2-monthly intervals. Lou/c rats have a high incidence of spontaneous myeloma production. For human hybridomas, [rodent myeloma × human lymphocytes] hetero-fusions were employed earlier but without consistent success. Therefore, tumour lines of human origin have been devised as a fusion partner. These include either myelomas or more frequently, lymphoblastoid cell lines (LCL) which are derived from EBV-transformed B-lymphocytes.

FEEDER CELLS

A number of culture systems require the use of feeder cells for growth. Depending upon the particular application, both normal cells obtained from different organs/tissues (thymus, spleen, bone marrow) and cell lines (3T3 fibroblasts, Epstein–Barr virus-transformed B-cell lines) are used. One of the main reasons for the use of feeder cells is to raise the cell density in limiting-dilution cultures for analysis of precursor cell frequencies or for cloning. Additional function of feeder cells are to provide accessory cell function, including both cell contact and secretion of growth factors or cytokines; to provide a source of antigen-presenting cells; and, in some systems, to provide a stimulation (antigen) for T cells (i.e., allogeneic feeder cells). Depending on the particular system, the growth of feeder cells may have to be prevented by exposure to γ-irradiation or mitomycin C. The use and applications of feeder cells are given in Table 11.2.

Table 11.2 Use and applications of feeder cells

Feeder cell	Cell type	Application	Comments
Thymocytes	Mostly T cells	LDA, expansion and cloning of hybridomas	Source of IL-6. Thymocyte-conditioned medium or alternative sources of IL-6 can sometimes be used as substitutes.
Peritoneal cells	Macrophages/lymphocytes	LDA, cloning of hybridomas, APC	Source of IL-1, IL-6; use as AC requires inactivation.
Spleen cells	Lymphocytes/macrophages	LDA, cloning of hybridomas, APC	Source of IL-1, IL-6; use as AC requires inactivation.
Bone marrow	Bone marrow stromal cells	Growth of pre-B cell	Source of IL-7.
NIH 3T3	Fibroblasts	LDA, growth of pre-B cells	-
EBV-transformed lymphoblastoid cell lines	Transformed B cells	LDA, cloning, culture of T cells and NK cells	Potential source of IL-12.

FUSION

Several fusion methods are employed to fuse together myeloma cells with normal B lymphocytes. The methods fall into one of the categories—biological, chemical and physical— as given in Table 11.3.Among them, the chemical method, which uses PEG as a fusogen, is the most common. A typical frequency for hybridoma generation in rodent fusion using PEG is between 8 and 10 hybridomas per 10^5 myelomas, whereas these frequencies are much lower in human systems, approximately, one hybridoma per 10^7 myelomas. Electrofusion which yields a higher fusion frequency than PEG, is favourable for human hybridomas.

Table 11.3 Fusion methods used in hybridoma production

Method	Details
Biological	Several groups of viruses such as HVJ (sendai virus) Rarely used now. Disadvantages: possible contamination of viral products.
Chemical	Polyethylene glycol (PEG) Cell membranes are fused together indiscriminately.
Physical	Electrofusion Cells are brought together in a low-amplitude continuous AC electric field, and then subjected to a high-amplitude short pulse. This brief pulse causes membrane breakdown, and cell fusion results when the membrane breakdown occurs in areas of cell–cell contact. High fusion frequencies: 10 times more efficient than PEG. Useful for human hybridomas Needs a speed device.

Several myeloma cell lines are available. A common mouse cell line used for the production of mAbs is X63-Ag8.653. The parental cell line was initially derived from a tumour induced by injecting mineral oil into a Balb/c mouse. The resulting cell line was subjected to mutagenesis to produce a cell line deficient in hypoxanthine-guanine phosphoribosyl transferase (HGPRT). Further cloning yielded X63-Ag8.653 which no longer secretes its own antibody. Therefore hybridomas prepared with this cell line will only secrete antibodies obtained from the spleen cells.

A fusogenic agent is added to the mixed myeloma and spleen cells. Polyethylene glycol (PEG) is the most common agent used for the fusion of cells. Other fusogenic agents will work, though. The spleen cells and myeloma cells are mixed and co-pelleted by centrifugation. A small volume of PEG solution in media is added and the cell pellet is gently dislodged. PEG fuses the plasma membranes of adjacent cells resulting in a heterokaryon. PEG is toxic and exposure needs to be minimized. The cells are washed to remove the PEG, resuspended in HAT medium, and then generally plated in 24-well plates.

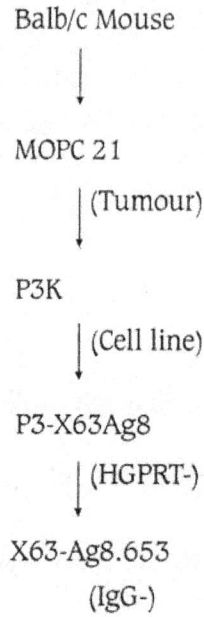

Balb/c Mouse

|

MOPC 21

| (Tumour)

P3K

| (Cell line)

P3-X63Ag8

| (HGPRT-)

X63-Ag8.653

(IgG-)

SELECTION OF HYBRID CELLS

HAT SELECTION

Not all of the cells will fuse following PEG treatment and many of the fusions will be homotypic. The unfused myeloma cells will tend to overgrow the fused cells in culture, and therefore, it is necessary to select for fused cells. The use of HGPRT-deficient myeloma cells is a commonly used genetic selection scheme for selecting fused cells. HGPRT is a crucial enzyme in the purine salvage pathway.

HAT MEDIUM

Selection of fused cells is achieved by adding hypoxanthine, aminopterin and thymidine (HAT) to the culture medium. Aminopterin is an inhibitor of dihydrofolate reductase (DHFR) which is a crucial enzyme for both de novo purine and pyrimidine syntheses. Hypoxanthine and thymidine are nucleotides utilized in the purine and pyrimidine salvage pathways, respectively. HGPRT-deficient myeloma cells will die in the presence of aminopterin since they are and cannot salvage purines and de novo purine synthesis is inhibited. Spleen cells contain an intact HGPRT and can therefore salvage, but do not have the ability to grow in culture. Therefore, only hybridomas (i.e., fusions between myelomas and spleen cells) are able to grow in the presence of HAT medium. The myeloma cells provide the ability to grow in culture and the spleen cells provide a functional HGPRT. The HAT medium is usually replaced with HT medium after 7–10 days after the fusion to minimize the aminopterin toxicity. Hypoxanthine and thymidine are still necessary, since aminopterin slowly decays and is still present in the medium. The plates are monitored for the next 2–4 weeks for the appearance of colonies.

HAT/OUABAIN DOUBLE SELECTION PROCEDURE

This method is used in human fusions where the antibody-producing cells to be fused are EBV-transformed lymphoblastocytes. The basis for the selection is shown in Table 11.4.

Table 11.4 Characteristics of different cells

Cells	HGPRT	Ouabain	Growth in HAT/Ouabain medium
Myeloma/lymphoblastoid cell line	–	Resistant	–
Antibody-producing lymphocyte	+	Sensitive	–
Hybridoma of the above two	+	Resistant	+

SCREENING

The goal of the screening step is to identify wells containing the mAb of desired specificity. Single cells secreting the desired antibody are then isolated from positive cultures and propagated into cell lines. Wells will sometimes contain mixtures of hybridomas. Thus a major issue in choosing a screening assay is that in addition to being fast and simpler, assays should have a clear-cut discrimination between positive and negative rather than exquisite sensitivity. In addition, some of the hybridomas will lose the ability to produce antibody due to chromosome loss or rearrangement.

CLONING

After identification of positive cultures, antibody-secreting hybridomas must be cloned to ensure that the antibody is homogeneous and monospecific. The term cloning in cell culture is defined as the initiation of a cell line from a single progenitor; that is, cloned cells are all genetically identical. The purpose of cloning is to not only select a desirable population but also to eliminate undesirable or unstable population. In hybridoma technology, it is necessary to repeat the cloning in order to ensure that non-producers, arising as spontaneous variants, do not overgrow the hybridoma of interest. (Cloned hybridoma cells must be expanded gradually from 96-well plates to 24-well plates and then to flasks. Large amount of hybridoma-derived antibodies can be prepared either *in vitro* or *in vivo* later).

Several methods are available for cloning cells and are listed in Table 11.5.

Table 11.5 Methods for cloning cells

Methods	Cells
Limiting dilution	Most cell types; often used for cloning T-cell clones and T- and B-hybridomas.
Cloning in semi-solid media	Hybridomas, tumour cells and bone marrow cells.
Micromanipulation (uses a cloning ring or cloning plate)	Adherent cells
Cell sorter (uses a continuous-flow cytofluorimeter	–

Two common methods utilized for cloning hybridomas are soft agar and limiting dilution.

LIMITING DILUTION METHOD

In the limiting dilution method, cells in the culture are enumerated, diluted and aliquot into new wells of 96 well-plate so that each well contains only one cell. Cells are allowed to regrow and the procedure is repeated many times to ensure that all cells in a given well are monoclonal. Cells are seeded into wells at very low cell densities, viz., 5 cells/ml or 10 cells/ml (100 µl/well of a 96-well plate), in the presence of appropriate concentration of antigen and 1×10^6 irradiated feeder cells. For hybridomas or T-cell lines, these may be mouse thymocytes, spleen cells or peritoneal exudate cells. The feeder cells must be irradiated at 30–40 GY or treated with mitomycin C to prevent any growth in culture, and need not be histocompatible. Positive cultures at the lowest cell density employed may be assumed to be clones, and are subjected to further analysis. Cloning should be carried out at least twice.

VERIFICATION OF CLONALITY

Depending on the cell type, several different methods have been used to verify the clonality of a culture.

Box 11.3 Methods to verify clonality	
Methods	Comments
Immunocytochemistry	Used to analyse a distinct cell-surface marker, e.g., a CD marker on lymphocytes. All cells should show the same genotype.
SDS-PAGE	Used to analyse a cell product to demonstrate homogeneity, e.g., immunoglobulin from a monoclonal hybridoma should demonstrate only 1–3 bands on SDS-PAGE according to different degrees of glycosylation.
Isoelectric focusing	Used to analyse a cell product as monoclonal antibody.
RFLP	Used to analyse a rearranged pattern of immunoglobulin or T-cell receptor gene.
Recloning	If cloning is carried out several times, the probability of the line being clonal is greatly increased.

SOFT AGAR METHOD

This exploits the fact that if many malignant cells will proliferate in a semi-solid medium containing low amounts of agar to form spherical colonies that will be spaced, then colonies picked out of agar are most likely to be monoclonal. For semi-solid culture, plate out cells are plated out at low densities into semi-solid media in which single cells grow into discrete colonies. Each colony is picked off using a sterile Pasteur pipette and transferred to a 96-well plate. Agar, agarose, or methyl cellulose can be used as medium.

In most cases, both methods are combined, since the clones require rescreening and are grown by using several cycles of limiting dilution. With both methods, it is recommended to subject the hybridoma to several rounds of cloning until a stable hybridoma is obtained. The medium from each well is screened using ELISA, and positive cultures are dispersed in soft agar. The individual colonies are then cultured in new wells and transferred to larger vessels successively until stocks can be frozen and enough material is available for characterization.

PRODUCTION OF MONOCLONAL ANTIBODY

Large quantities of mAbs can be produced by either *in vitro* culture or through the generation of ascites tumours. mAb produced *in vitro* is sometimes too dilute and will generally be contaminated with bovine serum components. Therefore, it may be necessary to purify and/or concentrate the mAb. Ascites are produced by injecting the hybridoma into the peritoneal cavity of a mouse. The hybridoma cells will produce an ascites tumour and secrete mAb into the peritoneal cavity. The ascites fluid is then collected several weeks later. Ascites can produce very high concentrations of antibody. The ascites fluid will be slightly contaminated with other mouse IgG.

CHARACTERIZATION AND STORAGE

Characterization is the process which establishes the monoclonality of the antibody. This, requires the biochemical and biophysical characterization of the antibody. Spectrometric, electrophoretic and chromatographic methods are used for this purpose. Suitability of the antibody for its intended use is also tested. For example, for therapeutic or diagnostic uses, the monoclonal antibody is coupled to a drug or radioisotopes, or enzyme; associated with a target cell population after injecting into an animal. The affinity of the antibody towards a particular antigen is also tested immunochemically. The antibody is also characterized for immunoglobulin class, the epitope for which it is specific, and the number of binding sites it possesses. The stability of antibody and the cell line is also an important aspect. The ability of the cell to withstand freezing, storage for various time periods and reculture must be ascertained. The physical and chemical stability of the antibody during storage and use for its intended purpose must be determined.

The cell lines secreting the antibody must be frozen in liquid nitrogen at several stages of cloning and culture to avoid the destruction of the clones. Seed stocks should also contain maximum number of clones secreting the antibodies of desired specificity, so that alternatives are available if the desired use of the antibody is changed or if the antibodies are proved to be unstable.

LARGE-SCALE PRODUCTION OF MONOCLONAL ANTIBODIES

A total of approximately around 50,000 mAbs are reported so far. Most are produced in small quantities. However, some have become commercially successful and so require a scale of production different from that usually experimented in research facilities. Commercial interests consider production scales of 0.1–10 g as small, 10–100 g as medium, and over 100 g as large. Monoclonal antibody production is generally performed for three purposes: diagnosis, therapy, and research and development of new therapeutic agents.

MONOCLONAL ANTIBODY PRODUCTION FOR DIAGNOSTIC AND THERAPEUTIC PURPOSES

The amount of mAb needed and the importance of factors such as cost, turn-around time, and regulatory compliance depend on the purpose of use. The very competitive diagnostic industry is concerned with cost, turn-around time, and regulatory requirements. The diagnostic–industry scale of mAb production is usually small to medium and seldom large. The therapeutic industry is considerably less concerned with cost and turn-around time than the diagnostic industry, and its production scale is medium to large. The therapeutic industry is highly regulated and sensitive to regulatory structure and to the very high regulatory cost of any procedural change.

Monoclonal antibody production requires more than the culturing of large batches of cells or their injection into large numbers of mice. It requires considerable pre-production effort to ensure that the cell line is stable, can produce commercially appropriate quantities of a stable antibody, and can produce an uncontaminated product. Commercial production also involves building a high-quality facility for *in vivo* and *in vitro* production and for processing of the antibody. There is a need for quality control and quality assurance departments to meet the requirements of good manufacturing practices that are required for commercial products. Product-lot testing is necessary to ensure product reproducibility. Production-process verification and documentation are necessary to protect the consumer.

IN VIVO AND *IN VITRO* METHODS FOR COMMERCIAL PRODUCTION OF MAb

Monoclonal antibody production uses both the mouse ascites method and *in vitro* methods. Cost is usually the major consideration in determining the method except for marketed therapeutic products. When fully-loaded production, and pre-production and post-production costs are considered for a commercially viable line, economics usually favour *in vivo* production. However, as the amount of mAb increases, existing *in vitro* production technology can become more economical because high, fixed optimization costs (costs associated with selecting a subclone with the best growth and mAb production characteristics and that grow in low-serum or serum-free conditions) associated with *in vitro* production are spread over a larger production amount, making cost per gram competitive with *in vivo* production, which has a higher and more variable cost structure. When production costs are compared for small-scale production, *in vitro* methods are 1/2 to 6 times higher, depending on the cell line. However, these costs might not include all factors such as animal housing costs and technician time. In large-scale production runs, *in vitro* systems are economically competitive and are usually selected because they reduce animal use and decrease the presence of contaminating foreign antigens if serum-free media can be used. When the time of mAb production is critical and small amounts are required, *in vivo* production is selected because it takes only 6 weeks.

The therapeutic industry uses primarily serum-free *in vitro* technology because of a concern for treatment-related allergic responses caused by repeated foreign-antigen exposure. Immune responses are of concern here because mice are the source of the cell lines used in most mAb production methods. The human immune system tends to reject mouse-derived antibodies, which can lead to allergies or decreased effectiveness of injected mAb. Therefore, techniques that replace most of the mouse's antibody genes with human DNA have been developed. Humanizing antibodies and producing antibody in SCID mice or in an *in vitro* system have alleviated this problem. In the therapeutic industry, early work to determine whether the mAb will have the desired effect is usually done with *in vivo*-derived mAb because turn-around time is shorter and production costs are lower. In the diagnostic industry, keen competition leads to over-riding cost considerations, whereas the presence of foreign antigens is less important. As a result, *in vivo*-derived products are commonly used. *In vivo* procedures are optimized to increase productivity by reducing hybridoma invasiveness and increasing mAb secretion. This optimization can result in a reduction in animal use by a factor of 2–10 that greatly reduces production costs. For very-small-scale production, ascites production is often used because it is a much more forgiving procedure than *in vitro* production and can be done without optimizing cell lines in an *in vitro* culture.

IN VIVO PRODUCTION

Biologic behaviour of a hybridoma cell line is very important in determining whether *in vitro* culture will be successful or the ascites method must be used. Biologic behaviour also affects the concentration of mAb produced and for the mouse ascites method, the quantity of ascites produced. Researchers, and production facility personnel can optimize production results of both *in vitro* and *in vivo* methods by adjusting production variables and selecting appropriate clones.

The variables affecting *in vivo* production and optimization include age, sex, strain of the host, size of the hybridoma-cell inoculum, number of taps, and type and volume of primer. Those variables can be manipulated to affect ascites yields and mAb concentration. For instance, low-ascites-producing subclones usually form only a few large tumours in the peritoneal cavity, whereas high-producing subclones form numerous colonies of small tumours that grow extensively throughout the mesentery. Therefore, clones that form less-invasive small soft tumours should be selected. Sequential tapping provides the highest yields and greatest mAb concentration from a group of mice. Except for very invasive cell lines that allow for only one needle tap, sequential tapping usually reduces the number of mice needed per gram of mAb by a factor of 2–3. The number of needle taps allowed should therefore be based on the clinical condition of the mice, and the maximum, in general, should be three taps.

Optimal *in vivo* production requires reduction of the invasive nature of a cell line so that all of the mice survive completion of a production run. Selecting appropriate clones and altering hybridoma cell concentration injected into the peritoneal cavity of the mice are two ways to optimize production. The volume and concentration of mAb produced depend on the clone selected, and this makes systematic comparisons difficult. Therefore, the best way to achieve maximal *in vivo* yields is to screen clones in mice and to use the clone that provides the best yield. Cell growth conditions are optimal *in vivo*, so almost all cell lines will produce antibody, even when they are not optimized. That is why injecting into mice usually saves cell lines that are difficult to grow *in vitro*.

Ascites production is a simple procedure, once proper technique is learned. Daily observation of the mice requires skilled observers to determine the optimal time for tapping the fluid and to determine when the mouse should be euthanized. It is quicker, is more forgiving, is more economical for small-scale and medium-scale production, produces a higher concentration of mAb, and is easier to scale up in production. For most cell lines, purification costs are the same as *in vitro* methods. The major problems associated with *in vivo* production are the use of animals, the presence of endogenous mouse immunoglobulin contamination except when immunodeficient mice are used, and the possibility of contamination with murine pathogens, which requires the use of high-quality animals and a high-quality program for health assurance. High-speed centrifugation of the ascitic fluid brings pristane to the top, where it can be removed easily.

IN VITRO PRODUCTION

Numerous *in vitro* commercial systems meet the different needs and requirements of users. These systems are of two types: single-compartment systems that allow only low-density cell culture and double-compartment systems that allow high-density cell culture, which results in increased mAb concentration. For very-small-scale production (less than 10 g), the simple low-density cell-culture systems-such as culture flasks, roller bottles, gas-permeable bags, and hollow-fibre bioreactors are used. For small-scale and medium-scale production (10–100 g), double-compartment, high-density cell-culture systems such as hollow-fibre systems, are used, as well as spinner flasks and roller bottles. High-scale production (over 100 g) is performed in large capital-intensive system, such as homogeneous suspension culture in deep-tank stirred fermenters, perfusion-tank systems, air-lift reactors, and continuous-culture systems.

An antigen-free product can be obtained by adapting the cell line to low-serum or serum-free media, with generally minor inhibitory effects on the cell line. Benefits of *in vitro* production are the absence of live-animal use, although some products in the culture media come from animals; the possibility of low-serum or serum-free media production; and the absence of host-contributed immunoglobulin or antigens. The problems associated with *in vitro* systems are:

- material, labour, and equipment costs are higher than for the *in vivo* method; characteristics of the hybridoma are more critical than *in vivo*;
- about 3–5% of all clones cannot be maintained in existing *in vitro* systems;
- the great potential for microbial contamination, poor growth, and mechanical failure of the system or supporting systems requires constant monitoring and attention every day;
- production of large quantities of mAb is slower because of low mAb concentration, compared with the ascites method;
- the increased employee technical capabilities and educational background required by increased training time and system manipulations increase labour expense;

- the design of downstream processing is emphasized because large volumes of media are required to obtain large quantities of mAb and to ensure product economy and purity;

- residual endotoxin, residual DNA from cell death, and bovine IgG contamination with cell lines that require some serum all complicate the process.

Advantages of in vitro method

- *In vitro* methods reduce the use of mice at the antibody-production stage but can use mice as a source of feeder cells.

- *In vitro* methods are usually the methods of choice for large-scale production by the pharmaceutical industry because of the ease of culture for production, compared with use of animals, and because of economic considerations.

- *In vitro* methods avoid the need to submit animal protocols to ethical committees.

- *In vitro* methods avoid or decrease the need for laboratory personnel experienced in animal handling.

- *In vitro* methods using semipermeable-membrane-based systems produce mAb in concentrations often as high as those found in ascitic fluid and are free of mouse ascitic fluid contaminants.

Disadvantages of in vitro methods

- It should be noted that each of the items below pertains to only a fraction (3–5%) of hybridomas, but they indicate some of the difficulties associated with *in vitro* methods.

- Some hybridomas do not grow well in culture or are lost in culture.

- *In vitro* methods generally require the use of FBS, which limits some antibody uses because it is a concern from the animal-welfare perspective.

- The loss of proper glycosylation of the antibody (in contrast with *in vivo* production) might make the antibody product unsuitable for *in vivo* experiments because of increased immunogenicity, reduced binding affinity, changes in biological functions, or accelerated clearance *in vivo*.

- In general, batch-culture supernatants contain less mAb (typically 0.002–0.01) per millilitre of medium than the mouse ascites method. Note that semipermeable-membrane-based systems have been developed that can produce concentrations of mAb comparable with concentrations observed in mouse ascites fluid.

- In batch tissue-culture methods, mAb concentration tends to be low in the supernatant; this necessitates concentrating steps that can change antibody affinity, denature the antibody, and add time and expense. Adequate concentrations of mAb might be obtained in semipermeable-membrane-based systems.

- Most batches of mAb produced by membrane-based *in vitro* methods are contaminated with dead hybridoma cells and dead hybridoma cell products, thus requiring early and expensive purification before study.

- mAb produced *in vitro* might yield poorer binding affinity than those obtained by the ascites method.

- *In vitro* culture methods are generally more expensive than the ascites method for small-scale or medium-scale production of mAb.

- The number of mAb produced by *in vitro* methods is limited.

Advantages of mouse ascites method

- The mouse ascites method usually produces very high mAb concentrations that often do not require further concentration procedures that can denature antibody and decrease effectiveness.

- The high concentration of the desired mAb in mouse ascites fluid avoids the effects of contaminants in *in vitro* batch-culture fluid when comparable quantities of mAb are used.

- The mouse ascites method avoids the need to teach the antibody producer tissue-culture methods.

Disadvantages of mouse ascites method

- The mouse ascites method involves the continued use of mice requiring daily observation.

- MAb produced by *in vivo* methods can contain various mouse proteins and other contaminants that might require purification.

- The mouse ascites method can be expensive if immunodeficient mice in a barrier facility must be used.

- *In vivo* methods can cause significant pain or distress in mice which may be a inhumane method.

APPLICATIONS OF MONOCLONAL ANTIBODIES

Hybridoma technology is being applied in diversified fields. The best example of the application of monoclonal antibodies is the generation of a large panel of reagents that define cell-surface structures on the surface of bone-marrow-derived cells that give rise to and make up our protective immune system. Monoclonal antibodies that define CD molecules and flow cytometry have been used to identify the normal components of the immune system and to determine if these components are under-represented (in the case of immunodeficiency diseases) or over-produced (in some cancers). It is well known that a subset of lymphocyte termed T-helper cells is important for normal immune function.

Monoclonal antibodies reactive with the CD4 molecule expressed on helper cells was used to demonstrate that a decrease in CD4 cells is a feature of AIDS and the levels can be used to stage the disease.Monoclonal antibodies have also been extensively used in the design of sensitive detection assays such as ELISAs. These tests have been used to detect normal and autoantibody levels, determine the presence and levels of autoantigens, viral/bacterial and other environmental antigens as well as assess the levels of normal components in bodily fluids.

Another area that monoclonals have had an impact on is the isolation and purification of molecules. A given monoclonal antibody can be coupled to an insoluble surface and used to affinity-purify the molecule of interest; this approach allows one to accomplish severalfold purification in a single step.

In molecular genomics a known monoclonal antibody that recognizes a molecule of interest can be used to identify its gene. Alternatively, if one has a newly identified gene with an unknown function, monoclonal antibodies can be generated against the predicted protein that it would encode and these reagents can be used for expression and function studies. These avenues also open up new windows of investigative research. For example, one can use monoclonal antibodies to determine if the gene is abnormally expressed in certain disease states or has a different structure in different individuals.

Monoclonal antibodies have been included in many protocols as biological response modifiers. Antibodies against bioactive cytokines have been used in therapy for many immunologically based disease processes such as the control of transplantation rejection and to modulate auto-immune diseases. Modifying monoclonal antibodies so as to be used as immunotoxins which can be used to search out and destroy a specific cell type such as cells expressing a specific tumour marker that the antibody recognizes. This mechanism is used to attack certain tumours as well as being used by some to manipulate an immune response.

Because one has isolated a monoclonal antibody one can easily clone the immunoglobulin genes that encode it. As many monoclonal antibodies are generated in rodent species,the use of such reagents in humans is limited since the immune system can recognize the rodent antibody as foreign and mount a potent response against it. As a result there is considerable interest to "humanize" these antibodies by replacing those rodent structures with human counterparts and perhaps allow the reagent to be ignored by the human immune system.

Monoclonal antibodies have also been used in applied chemistry. It was appreciated that enzymes function in part by having a high-affinity interaction with a short-lived transition state. It was reasoned that, if enzymes can interact with these transitional states, maybe an antibody can as well and potentially serve as an enzyme. By generating antibody reagents against enzyme inhibitors that mimic the transitional state, one can isolate antibodies that can provide catalytic function. The catalytic reactions range from redox reactions to structural rearrangements. This initial success, together with the principles emerging in the field of combinatorial chemistry, indicates that recognition molecules generated by the immune system have tremendous potential to be used as chemical tools.

APPLICATIONS OF MONOCLONAL ANTIBODY IN VARIOUS FIELDS OF MEDICAL SCIENCE

1. Dose determination of a medicine can be carried out by using the monoclonal antibodies of an animal, immunized against this particular medicine.

2. They are used

   - to detect allergies,
   - to carry out hormone tests,
   - to diagnose viral diseases,
   - to detect certain types of cancers,
   - to monitor the presence or appearance of malignant cells after surgical or radio-therapeutic treatments.

3. The purification of complex mixtures or substances, the biological role of which is important (proteins, hormones, toxins, etc.), could also be carried out with monoclonal antibodies.

4. The use of these antibodies was also envisaged for the labelling and precise identification of specialized cells such as neurons in order to gain better knowledge of the way in which these cells associate and operate.

5. Monoclonal antibody technique is also of great value in the area of the structure of cell membrane as membrane proteins are hard to purify.

6. In the field of direct therapy, serotherapy can be made more effective with the administration of a monoclonal antibody.

7. Monoclonal antibodies could also be used in the preparation of very specific vaccines, particularly against certain viral strains and against other parasites.

8. Monoclonal antibodies could also neutralize the action of lymphocytes responsible for the rejection of grafts, and destroy the auto-antibodies produced in auto-immune diseases.

9. In association with medicinal substances, they could considerably increase the effectiveness of the latter on the target cells, while avoiding the serious side-effects of cancer therapies.

10. Certain companies are preparing diagnostic kits designed for the screening of certain lethal diseases. It is anticipated that the future support of some aspects of this hybridoma-

based monoclonal antibody technology will be tailored to the needs of each developing country in the world.

MAb for Cancer Research

Monoclonal antibodies, in addition to being highly effective detectors of cancer cells, are under consideration as potential means of tumour therapy.

Applications of monoclonal antibodies in a nutshell

- *Diagnostic applications*

    Biosensors

    Microarrays

- *Therapeutic applications*

    Transplant rejection, e.g., Muronomab-CD3

    Cardiovascular disease, e.g., Abciximab

    Cancer, e.g., Rituximab

    Infectious diseases, e.g., Palivizumab

    Inflammatory disease, e.g., Infliximab

- *Clinical applications*

    Purification of drugs

    Imaging the target

- *Future applications*

    Fight against bioterrorism

The possible roles for monoclonal antibodies include:

- the delivery of high concentration of cell-killing radioactive chemicals, drugs, or toxins to cancerous tissues and
- as well as the use of these antibodies to locate and attack tumour cells.

For effective delivery of anticancer agents linked to monoclonal antibodies, the combination (conjugates) should penetrate and destroy the parts of a tumour that contribute to its growth and then bind directly to most, if not all, individual tumour cells. Additionally, anticancer drugs may be released around the tumour cells (extracellularly) following monoclonal antibody attachment to tumour cell surfaces.

MAJOR COMPANIES MAKING MONOCLONAL ANTIBODIES, AND THEIR INDICATIONS

Company name	Name of product	Indications
Ortho Biotech	Orthoclone-OKT®	Organ transplant rejection
J & J/Eli Lilly	ReoPro®	Acute cardiac conditions
BiogenIdec/Genentech/Roche	Rituxan®	Non-Hodgkin's lymphoma
BiogenIdec	Zevalin™	
PDLI	Zenapax®	Acute transplant rejection
MedImmune/Abbott	Synagis®	Viral respiratory disease
Genentech/Roche	Herceptin®	Breast cancer
	Avastin®	Colorectal cancer
J & J	Remicade®	Crohn's, rheumatoid arthritis
Novartis	Simulect®	Acute myeloid leukaemia
Wyeth	Mylotarg™	Acute myeloid leukaemia
Schering/ILEX Oncology	Campath®	Chronic lymphocytic leukaemia
Abbott/CAT	Humira™	Rheumatoid arthritis
Novartis/Genentech/Tanox	Xolair®	Asthma
Genentech/Xoma	Raptiva™	Psoriasis
Corixa/GlaxoSmithKline	Bexxar®	Non-Hodgkin's lymphoma
BMS/ImClone Systems	Erbitux™	Colorectal cancer

REVIEW QUESTIONS

1. Write in detail the various steps involved in production of monoclonal antibody.

2. What are the different applications of monoclonal antibodies?

12

SOMATIC CELL NUCLEAR TRANSFER

In genetics and developmental biology, somatic cell nuclear transfer (SCNT) is a laboratory technique for creating an ovum with a donor nucleus (*See* process below). It can be used in Embryonic stem cell research, or in regenerative medicine where it is sometimes referred to as "Therapeutic cloning." It can also be used as the first step in the process of reproductive cloning.

THE PROCESS

A narrow tube, probably a section from a micropipette, removes the maternal chromosomes from an oocyte prior to the somatic cell nuclear transfer. In SCNT, the nucleus of a somatic cell (a body cell other than a sperm or egg cell), which contains the organism's DNA, is removed and the rest of the cell discarded. At the same time, the nucleus of an egg cell is removed. The nucleus of the somatic cell is then inserted into the enucleated egg cell. After being inserted into the egg, the somatic cell nucleus is reprogrammed by the host cell. The egg, now containing the nucleus of a somatic cell, is stimulated with an electric shock and will begin to divide. After many mitotic divisions in culture, this single cell forms a blastocyst (an early-stage embryo having about 100 cells) with DNA almost identical to the original organism.

LIMITATIONS

Both the egg cell and the introduced nucleus are put under enormous stress, leading to a high loss in resulting cells. For example, Dolly the sheep was born after 277 eggs were used for SCNT, which created 29 viable embryos. Only three of these embryos survived until birth, and only one survived to adulthood. As the procedure currently cannot be automated, but has to be performed manually under a microscope, SCNT is very resource-intensive. The biochemistry involved in reprogramming the differentiated somatic cell nucleus and activating the recipient egg is also far from being understood.

In SCNT, not all of the donor cell's genetic information is transferred, as the donor cell's mitochondria which contain their own mitochondrial DNA are left behind. The resulting hybrid cells retain those mitochondrial structures which originally belonged to the egg. As a consequence, clones such as Dolly that are born from SCNT are not perfect copies of the donor of the nucleus.

CONTROVERSY

Proposals to use nucleus transfer techniques in human stem cell research raise a set of concerns beyond the moral status of any created embryo. These led to some individuals and organizations who are also opposed to human embryonic stem cell research to be concerned about, SCNT research.

- Blastula creation in human stem cell research will lead to the reproductive cloning of humans. Both processes use the same first step—the creation of a nuclear transferred embryo, most likely via SCNT. Those who hold this concern often advocate for strong regulation of SCNT in the usage of derived products for the intention of human reproduction, or its prohibition.

- The appropriate source of the eggs that are needed for this procedure SCNT requires human eggs, which can only be obtained from women. The most common source of these eggs today are those unused eggs produced in excess of the clinical need during IVF treatment. This is a minimally invasive procedure, but it does carry some health risks, such as ovarian hyperstimulation syndrome, may increase the risk of ovarian cancer and in very rare instances even death.

POLICIES

SCNT is currently legal for research purposes in the United Kingdom, having been incorporated into the 1990 Human Fertilization and Embryology Act in 2001. Permission must be obtained from the Human Fertilisation and Embryology Authority in order to perform or attempt SCNT. In the United States, the practice remains legal, as it has not been addressed by federal law. In 2005, the United Nations adopted a proposal submitted by Costa Rica, calling on member states to "prohibit all forms of human cloning in as much as they are incompatible with human dignity and the protection of human life." This phrase may include SCNT, depending on interpretation.The Council of Europe's Convention on Human Rights and Biomedicine and its Additional Protocol to the Convention for the Protection of Human Rights and Dignity of the Human Being with regard to the Application of Biology and Medicine, on the Prohibition of Cloning Human Being appear to ban SCNT. Of the Council's 45 member states, the Convention has been signed by 31 and ratified by 18. The Additional Protocol has been signed by 29 member nations and ratified by 14. In India strict rules are being paved by the Government to streamline the issue.

The following is the list of animals that have been cloned in chronological order.

1. Tadpoles	11. Cat
2. Humans	12. Rabbit
3. Carp	13. Mule
4. Mice	14. Deer
5. Sheep	15. Horse
6. Rhesus Monkey	16. Rat
7. Pig	17. Fruit flies
8. Gaur	18. Dog
9. Cattle	19. Wolf
10. Dairy cattle	

ANIMAL CLONING

Transgenic animals (animals engineered to carry genes from other organisms, usually of species other than their own) can be made to produce a wide variety of proteins that could be sold as drugs, as well as enzymes that could be used to speed up industrial chemical reactions. Although the creation of transgenic animals began in the 1980s, cloning was expected to make it possible for mass production of such animals. Large numbers of transgenic animals could produce vast quantities of drugs and other substances more efficiently at lower cost than is possible with such bioengineering technology such as bioreactors, steel vessels in which billions of genetically modified microorganisms produce proteins that are then extracted and purified.

Researchers involved in cloning envision a number of other practical applications for their work, such as the creation of genetically modified animals that could provide organs suitable for human organ transplants, the mass production of faster-growing and leaner strains of livestock free of disease causing genes and the perpetuation of endangered species.

HUMAN CLONING

The same procedures used to clone sheep and cattle could theoretically be used to clone humans. However, human cloning would probably be more difficult than sheep or cattle cloning, because the cells of human embryos start producing proteins at a relatively early stage. Thus, there would not be enough time for the egg cytoplasm to reprogram a transplanted nucleus. However, the successful (1998) cloning of mice, which also start producing proteins at an early embryonic stage, strongly indicated that this problem can be overcome in humans.

Infertile couples who do not wish to adopt, for instance, could use cloning to have children who are biologically related to them. Cloning could also be used to produce offspring free of certain diseases. For example, some disorders affecting the eyes, brain, and muscles, are (at least partially) caused by flawed genes located in the mitochondria—energy-producing structures in the cytoplasm. If a woman carrying a gene for one of these disorders, could conceive a healthy child by having the nucleus of one of her body cells inserted into an enucleated egg cell from a woman who has normal mitochondrial genes. The resulting embryo could then be implanted into the woman who donated the nucleus, who would carry the baby to term.

Some Proposed Practical Applications of Cloning

- The mass production of animals genetically engineered to carry human genes for the production of certain specific proteins that could be used as drugs; the proteins would be extracted from the animals' milk and used to treat human diseases.
- The mass production of animals with genetically modified organs that could be safely transplanted into humans.
- The mass production of livestock that have been genetically modified to possess certain desirable traits.
- The perpetuation of endangered species.
- The production of offspring by infertile couples.
- The production of an offspring free of a potentially disease causing genetic flaw carried by one member of the couple; the resulting individual without the defect could be cloned.

Dolly and Other Cloning Breakthroughs Since 1996

In 1996, however, researchers led by embryologist Ian Wilmut of the Roslin Institute near Edinburgh, Scotland, took mammary gland cells from an adult sheep and placed them in a solution that essentially starved them of nutrients and caused them to stop growing for a few days. Then, with a spark of electricity, they fused each mammary cell with an enucleated egg cell. The resulting cells were allowed to develop into embryos, which were then transplanted into surrogate mother *Ewes* (female sheep) to complete their development. Nearly 300 attempts at this technique resulted in failure for the scientists. Some eggs did not accept the mammary cell nuclei, embryos that were produced died, and lambs that were born were abnormal and died. But one lamb—Dolly—that survived the procedure and was apparently healthy was born in July 1996. The most important factor is that the nucleus of a specialized body cell from an adult animal could be reprogrammed to direct the development of a new organism, that is, the cell could be restored to its totipotency (the capability, under certain circumstances, of directing the development of a complete organism). The cell-starvation technique used by the scientists was based on the theory that it is easier to reprogram cells that are not growing. The programming of a nucleus is an interactive process between the cytoplasm (the cellular substance surrounding the nucleus) and the genes in the nucleus.

The cytoplasm sends to the nucleus signals that determine which genes are turned on or off and therefore which proteins—the end product of genes—are produced by the cell. As the cell becomes specialized and loses its totipotency, it becomes unable to produce many proteins.

In nuclear transplantation, the longer the cytoplasm has to work on the transplanted nucleus before protein synthesis begins, it is more likely that the nucleus can be reprogrammed to direct the normal development of an embryo.

TRANSGENIC ANIMALS

In July 1997, Ian Wilmut and his colleagues announced that they had combined their cloning procedure with genetic engineering techniques to produce lambs with a human gene for a particular blood protein in their cells. First, Wilmut's team inserted the human gene into a sheep foetus cell. This cell was then allowed to divide into many cells, each of which was then used in a nuclear transplantation to produce an embryo. Of the three identical female lambs born with the human gene, two—named Polly and Molly—survived. Animals engineered to carry genes from other species are called transgenic animals. The human gene that Polly and Molly were born with enables them to produce factor IX, a human blood-clotting protein useful for treating haemophilia, a disorder in which blood clotting does not occur normally. The protein is secreted into the animals' milk and can be extracted to create a drug for treating haemophilia. In January 1998, scientists at the University of Massachusetts and Advanced Cell Technology, incorporated, a biotechnology firm in Worcester, Massachusetts, announced the birth of three transgenic calves created through a method similar to that used with Polly and Molly. Although these calves carried only an experimental gene that had no effect on the animals, the scientists said that other calves they planned on creating were to carry a gene to produce human serum albumin, a protein needed by people who have lost large amounts of blood.

HAWAII MOUSE CLONING

The Hawaii researchers used naturally dormant **cumulus cells** (cells that surround eggs in ovaries) in cloning procedure. Because these cells were not growing, they could be easily reprogrammed inside enucleated egg cells without starving them in a special solution, as was necessary with the udder cells used in the Dolly procedure. And instead of electrically fusing a body cell with an enucleated egg cell, as was done in the Dolly technique, the Hawaii researchers used an extra-fine needle to inject the nucleus from a cumulus cell into an enucleated egg cell. Because this technique did less damage to the egg than did electrical fusion, it increased the chance of the resulting cell to develop into a healthy embryo. Scientists state that the success of this mouse research technique makes it more likely that humans could be successfully cloned since there are important similarities between mouse and human development.

First Clones of Pigs

The first clones of an adult pig were born on March 5, 2000. Scientists at PPL Therapeutics, Inc., in Edinburgh, Scotland, created the five pig clones using genetic material taken from a body cell of an adult female pig. The researchers said, in the future, clones of genetically modified pigs might serve as sources of organs for human organ transplants.

In April 2001, PPL Therapeutics announced that it had created a group of genetically engineered pigs through the cloning procedure. The pigs contained an easy-to-identify gene from a jellyfish that scientists added to the clones' genetic code as an experiment. They said the experiment was a step toward genetically manipulating pigs so that their organs would not be rejected by the human body.

Endangered Species Cloned

The first clone of an endangered species was born on Jan 8, 2001. However, the young animal—a type of wild ox called a gaur—died just two days later from a common bacterial infection. Gaurs are native to India, Myanmar, and the Malay Peninsula. The animals were on the verge of extinction in 2001 because of hunting and habitat destruction. Some conservationists have advocated cloning as a way to increase the populations of severely endangered species like the gaur.

To clone the gaur, researchers used hundreds of skin cells that had been removed from a deceased adult gaur. They extracted the DNA-containing nucleus from each of the cells and implanted the nuclei into bovine (cow) egg cells, from which the cow nuclei had been removed. Of nearly 700 cow eggs given gaur DNA, about 80 developed as embryos (organisms in an early stage of development), 44 of which were transplanted into 32 surrogate-mother cows. Eight of the cows became pregnant, but only one gave birth to a gaur that the scientists named Noah. Noah was genetically identical to the adult from which the skin cells had been obtained.

Review Questions

1. What are the practical applications of cloning?
2. Write short notes on:
   i. Dolly
   ii. Transgenic animals

13

GENETIC MANIPULATION OF ANIMALS

A new era in animal research was ushered in during the early 1980s when successful experiments designed to genetically modify animals by inserting foreign DNA were first reported. These new methodologies were used for

- *Identification of gene function*—The ability to insert genes into whole animals or to selectively delete or alter single predetermined genes in an animal provides enormous power in studying gene function.

- *To study animal models of disease*—Through altering even single genes within a living animal in such a way as to mimic mutations faithfully in an analogous gene in humans, thereby providing a higher chance of resembling human disease phenotypes.

In order to create genetically modified animals, it is necessary to modify the DNA of germline cells so that the modified DNA is heritable. As a result, certain cells that have the capacity to differentiate into the different cells of an adult animal were considered to be the optimal targets for introducing foreign DNA. The fertilized oocyte is one such cell, being totipotent. Other target cells are cells of very early stage embryos, including **embryonic stem (ES) cells**. Although such cells represent a stage in development where there has been incomplete separation of the somatic and the germline, such cells are therefore capable of giving rise to both somatic and germline cells.

When a foreign DNA molecule is artifically introduced into the cells of an animal, a transgenic animal is produced. The foreign DNA molecule is called a transgene and may contain one or many genes. By inserting a transgene into a fertilized oocyte or cells from the early embryo, the resulting transgenic animal may be able to transmit the foreign DNA stably in its germline. Many different types of transgenic animals have been created including transgenic *Drosophila*, transgenic frogs, transgenic fish and a variety of transgenic mammals including mice, rats and various livestock animals.

Transgenesis involves transfer of foreign DNA into totipotent or pluripotent embryo cells (either fertilized oocytes, cells of the very early embryo or cultured embryonic stem cells) followed by insertion of the transferred DNA into host chromosomes. If the foreign DNA integrates into the chromosomes of a fertilized oocyte, the developing animal will be fully transgenic since all nucleated cells in the animal should contain the transgene. If chromosomal integration occurs later, at a post-zygotic stage, the animal will be a mosaic, with some cells containing the transgene and some others lacking it. If the transgene is present in germline cells it can be passed through sperm or egg cells into some of the animal's progeny.

Although transgenes often integrate into the host chromosomes without affecting the expression of any endogenous genes, occasionally the integration event alters endogenous gene expression (insertional mutation), producing a recognizable phenotype. This constitutes a form of *in vivo* mutagenesis, albeit at an **unselected** target gene. Gene targeting was developed as a method of *in vivo* mutagenesis in which the mutation is introduced into a **preselected** endogenous gene. This can be achieved in somatic cells, but gene targeting in cultured ES cells is particularly powerful because it can lead to the construction of an animal in which all nucleated cells contain a mutation at the desired locus.

PRONUCLEAR MICROINJECTION

Fertilized oocytes are recovered from excised oviducts. The DNA of interest is then microinjected using a micromanipulator into the pronucleus of individual oocytes. Surviving oocytes are reimplanted into the oviducts of foster females and allowed to develop into mature animals.

During this procedure, the microinjected DNA (transgene) randomly integrates into chromosomal DNA, usually at a single site, although rarely two sites of integration are found in a single animal. As a result of chromosomal integration, the transgenes can be passed on to subsequent generations in mendelian fashion—if the foreign DNA has integrated at the one-cell stage, it should be transmitted to 50% of the offspring.

The microinjection of foreign DNA into fertilized oocytes is technically difficult and not suited to large-scale production of transgenic animals or to sophisticated genetic manipulation. An alternative involves transferring the foreign DNA initially into cultured **embryonic stem (ES) cells**. Mouse ES cells are derived from 3.5 to 4.5-day post-coitum embryos and arise from the **inner cell mass** of the blastocyst. The ES cells can be cultured *in vitro* and retain the potential to contribute extensively to all of the tissues of a mouse, **including the germline,** when injected back into a host blastocyst and reimplanted in a pseudopregnant mouse.

The developing embryo contains two populations of cells derived from different zygotes, those of the blastocyst and the implanted ES cells. If the two strains of cells are derived from mice with different coat colours, chimeric offspring can easily be identified. Use of genetically modified ES cells results in a partially transgenic mouse. Because the injected ES cells can form all or part of the functional germ cells of the chimera, it is possible to derive fully transgenic mice. This is usually accomplished by screening the offspring of matings between chimeras (usually males) and mice with a coat colour recessive to that of the strain from which the ES cells were derived.

The big advantage of ES cells is that they can be grown readily in culture. This means that a variety of genetic manipulations can be conducted in cultured ES cells. Importantly, **the desired genetic modification can be verified in tissue culture,** before injecting the genetically modified cells into a blastocyst prior to implantation. For example, the desired gene can be ligated to a marker gene, such as the *neo* gene, enabling a positive selection for cells that have been successfully transfected. The presence of the desired gene can also be verified quickly by a PCR-based assay. ES cells has the advantage of **gene targeting** by homologous recombination, a method which permits programmed selective alteration of a single **predetermined gene** removal of exons, and introduction of point mutations. Thereby gene targeting helps to identify gene function and also for creating animal models of human diseases.

SIGNIFICANCE OF TRANSGENIC ANIMALS

Transgenic animals have been extremely important for analysing human genes, and have helped greatly in our understanding of a variety of fundamental biological processes, notably in immunology, neurobiology, cancer and developmental studies. Some major applications are:

- **Identification of gene expression and its regulation** Although evidence for *cis*-acting regulatory elements is often inferred initially from studies using cultured cells, they need to be validated in whole animal studies. Transgenes consisting of the presumptive regulatory sequence(s) coupled to a reporter gene such as *lacZ*, provide a sensitive method of detecting gene expression and a powerful way of investigating regulation of gene expression. Long-range control of gene expression is often investigated using YAC transgenes.

- **Identification of gene function by targeted gene inactivation** Specific genes can be inactivated by a gene targeting procedure to introduce a transgene into the target gene known as "insertional inactivation". The effect on the phenotype of creating a null mutation in the gene of interest can provide powerful clues to gene function.

- **Investigating dosage effects and ectopic expression** In some cases, valuable information can be gained by over-expressing a transgene or by expressing it ectopically (the transgene is coupled to a tissue-specific promoter which causes expression in cells, the phenotypic consequence may provide valuable clues to function).

* **Identifying gain of function**   Any mammalian gene that produces a dominant negative effect or gain of function can be investigated by introduction of an appropriate transgene. In some cases, this can provide proof of a suspected biological function. A classical example concerns the *Sry* gene. A variety of different genetic analyses had implicated this gene as a major male-determining gene but convincing proof was obtained using a transgenic mouse approach. The experiment consisted of transferring a cloned *Sry* gene into a fertilized 46,XX mouse oocyte. As a result of this artificial intervention, the resulting mouse, which nature had intended to be female, turned out to be male.

* **Cell lineage ablation**   Transgenes can be designed consisting of a tissue-specific promoter coupled to a sequence encoding a toxin, for example diphtheria toxin subunit A or ricin. When the promoter becomes active at the appropriate stage of tissue differentiation, the toxin is produced and kills the cells. Thus, certain cell lineages in the animal can be eliminated (cell ablation) and the phenotypic consequences monitored.

* **Modelling human disease**   Insertional inactivation is often used to model loss-of-function mutations whilst gain-of-function mutations can often be modelled by inserting a mutant transgene.

YAC transgenics   A major breakthrough in transgenic studies was the development of YAC transgenics. The transfer of a 670-kb YAC containing the human HPRT (hypoxanthine phosphoribosyltransferase) gene into mouse ES cells. This was accomplished by spheroplast fusion (i.e., fusion of ES cells with YAC-containing yeast cells that have been stripped of the hard cell wall). Fragments from the yeast genome can integrate at the same time, however, and so alternative methods have sought to purify an individual YAC by size-fractionation on a preparative gel using pulsed-field gel electrophoresis (assuming that the YAC migrates at a position in the gel that is different from any yeast chromosome). The purified YAC can be inserted into a fertilized oocyte by pronuclear microinjection (*See* above). This method is, however, limited to small YACs, the DNA of large YACs is more likely to fragment following microinjection with very fine micropipettes.

TRANSCHROMOSOMIC ANIMALS

YACs have upper limits for the size of foreign inserts that can be transferred. Mammalian artificial chromosomes have also been generated, including first generation human artificial chromosomes. Such systems will have the capacity of transferring hundreds and possibly thousands of genes into transgenic animals, although a preferred route may be by using nuclear transfer technology rather than ES cells. Recently, however, transfer of whole chromosomes or chromosome fragments into ES cells has been made possible by microcell-mediated chromosome transfer. Using this approach transfer of human chromosomes or chromosome fragments derived from normal fibroblasts into mouse ES cells can be achieved. The resulting chimeric transchromosomic mice were viable, and the chromosome fragments appeared to show functional expression and could be transmitted through the germline.

Cells from the inner cell mass were cultured following excision of oviducts and isolation of blastocysts from a suitable mouse strain. Such embryonic stem (ES) cells retain the capacity to differentiate into, ultimately, the different types of tissues in the adult mouse. ES cells can be genetically modified while in culture by insertion of foreign DNA or by introducing a subtle mutation. The modified ES cells can then be injected into isolated blastocysts of another mouse strain (e.g., C57B10/J which has a black coat colour that is recessive to the agouti colour of the 129 strain) and then implanted into a pseudopregnant foster mother of the same strain as the blastocyst. Subsequent development of the introduced blastocyst results in a chimera containing two populations of cells (including germline cells) which ultimately derived from different zygotes (normally evident by the presence of differently coloured coat patches).

REVIEW QUESTIONS

1. Write short notes on:

   i. Microinjection

   ii. Transchromogenic animals

   iii. Significance of transgenic animals

14

STEM CELLS

INTRODUCTION

Stem cells are cells found in all multicellular organisms. They are characterized by the ability to renew themselves through mitotic cell division and differentiate into a diverse range of specialized cell types. In many tissues, stem cells serve as a sort of internal repair system, dividing essentially without limit to replenish other cells as long as the person or animal is still alive. Research in the stem cell field grew out of findings by Ernest A. McCullouch and James E. Till at the University of Toronto in the 1960s. When a stem cell divides, each new cell has the potential either to remain a stem cell or become another type of cell with a more specialized function, such as a muscle cell, a red blood cell or a brain cell. The two broad types of mammalian stem cells are: embryonic stem cells that are isolated from the inner cell mass of blastocysts, and adult stem cells that are found in adult tissues. In a developing embryo, stem cells can differentiate into all of the specialized embryonic tissues. In the 3- to 5-day-old embryo called a blastocyst, the inner cells give rise to the entire body of the organism, including all of the many specialized cell types and organs such as the heart, lung, skin, sperm, eggs and other tissues. In adult organisms, stem cells and progenitor cells act as a repair system for the body, replenishing specialized cells, but also maintain the normal turnover of regenerative organs, such as blood, skin or intestinal tissues. Stem cells can now be grown and transformed into specialized cells with characteristics consistent with cells of various tissues such as muscles or nerves through cell culture. Highly plastic adult stem cells from a variety of sources, including umbilical cord blood and bone marrow, are routinely used in medical therapies. Embryonic cell lines and autologous embryonic stem cells generated through therapeutic cloning have also been proposed as promising candidates for future therapies.

PROPERTIES OF STEM CELLS

All stem cells, regardless of their source, have three general properties—they are capable of dividing and renewing themselves for long period; they are unspecialized; and they can give rise to specialized cell types. The classical definition of a stem cell requires that it possesses two properties:

1. **Self-renewal**—the ability to go through numerous cycles of cell division while maintaining the undifferentiated state and

2. **Potency**—the capacity to differentiate into specialized cell types. In the strictest sense, this requires stem cells to be either totipotent or plutipotent, to be able to give rise to any mature cell type, although multipotent or unipotent progenitor cells are sometimes referred to as stem cells.

SELF-RENEWAL

Stem cells are capable of dividing and renewing themselves for long periods. Unlike muscle cells, blood cells or nerve cells, which do not normally replicate themselves, stem cells may replicate many times, or proliferate. A starting population of stem cells that proliferates for many months in the laboratory can yield millions of cells.

Two mechanisms exist to ensure that the stem cell population is maintained:

1. **Obligatory asymmetric replication** where a stem cell divides into one daughter cell that is   identical to the original stem cell, and another daughter cell that is differentiated.

2. **Stochastic differentiation** when one stem cell develops into two differentiated daughter cells, and another stem cell undergoes mitosis and produces two stem cells identical to the original.

The specific factors and conditions that allow stem cells to remain unspecialized are of great interest to scientists. It has taken scientists many years of trial and error to learn to derive and maintain stem cells in the laboratory without them spontaneously differentiating into specific cell types. For example, it took two decades to learn how to grow human embryonic stem cells in the laboratory following the development of conditions for growing mouse stem cells.

Stem cells are unspecialized. One of the fundamental properties of a stem cell is that it does not have any tissue-specific structures that allow it to perform specialized functions. For example, a stem cell cannot work with its neighbours to pump blood through the body (like a heart muscle cell), and it cannot carry oxygen molecules through the bloodstream (like a red blood cell). However, unspecialized stem cells can give rise to specialized cells, including heart muscle cells, blood cells or nerve cells.

POTENCY

Plutipotent, embryonic stem cells originate as inner mass cells within a blastocyst. The stem cells can become any tissue in the body, excluding a placenta. Only the morula's cells are totipotent, able to become all tissues and a placenta. Potency specifies the differentiation potential (the potential to differentiate into different cell types) of the stem cell.

    i. **Totipotent** (omnipotent) stem cells can differentiate into embryonic and extra-embryonic cell types. Such cells can construct a complete, viable, organism. These cells are produced from the fusion of an egg and sperm cell. Cells produced by the first few divisions of the fertilized eggs are also totipotent.

    ii. **Pluripotent** stem cells are the descendants of totipotent cells and can differentiate into nearly all cells. Examples include embryonic stem cells and cells that are derived from the mesoderm, endoderm and ectoderm germ layers that are formed in the beginning stages of embryonic stem cell differentiation.

    iii. **Multipotent** stem cells can differentiate into a number of cells, but only those of a closely related family of cells. Examples include haematopoietic (adult) stem cells that can become red and white blood cells or platelets.

    iv. **Oligopotent** stem cells can differentiate into only a few cells, such as lymphoid or myeloid stem cells.

    v. **Unipotent** cells can produce only one cell type, their own, but have the property of self-renewal which distinguishes them from non-stem cells (e.g., muscle stem cells).

Adult stem cells typically generate the cell types of the tissue in which they reside. For example, a blood-forming adult stem cell in the bone marrow normally gives rise to the many types of blood cells. It is generally accepted that a blood-forming cell in the bone marrow, which is called a haematopoietic stem cell, cannot give rise to the cells of a very different tissue, such as nerve cells in the brain.

IDENTIFICATION

The stem cell is a cell that has the potential to regenerate tissue over a lifetime. For example, the gold standard test for a bone marrow or haematopoietic stem cell (HSC) is the ability to transplant one cell and save an individual without HSCs. In this case, a stem cell must be able to produce new blood cells and immune cells over a long time, demonstrating potency. It should also be transplanted into another individual without HSCs, demonstrating that the stem cell was able to self-renew. Clonogenic assays (a laboratory procedure) can also be employed *in vitro* to test whether single cells can differentiate and self-renew. Researchers may also inspect cells under a microscope to see if they are healthy and undifferentiated or they may examine chromosomes. To test whether human embryonic stem cells are pluripotent, scientists allow the cells to differentiate spontaneously in cell culture, manipulate the cells so that they will differentiate to form specific cell types, or inject the cells into an immunosuppressed mouse to test for the formation of a teratoma (a benign tumour containing a mixture of differentiated cells).

EMBRYONIC STEM CELLS

Embryonic stem cell lines (ES cell lines) are cultures of cells derived from the epiblast tissue of the inner cell mass (ICM) of a blastocyst or earlier morula-stage embryos. A blastocyst is an early stage embryo—approximately 4–5 days old in humans and consisting of 50–150 cells. ES cells are pluripotent and give rise during development to all derivatives of the three primary germ layers, ectoderm, endoderm and mesoderm. In other words, they can develop into each of the more than 200 cell types of the adult body when given sufficient and necessary stimulation for a specific cell type. Nearly all research to date has taken place using mouse embryonic stem cells (mES) or human embryonic stem cells (hES). Both have the essential stem cell characteristics, yet they require very different environments in order to maintain an undifferentiated state. A human ES is also defined by the presence of several transcription factors and cell-surface proteins. The transcription factors Oct-4, Nanog and Sox2 form the core regulatory network that ensures the suppression of genes that lead to differentiation and the maintenance of pluripotency. The cell-surface antigens most commonly used to identify hES cells are the glycolipids SSEA3 and SSEA4 and the keratin sulphate antigens Tra-1-60 and Tra-1-81. Differentiation of mouse embryonic stem cells is shown in Figure 14.1.

Laboratories that grow human embryonic stem cell lines use several kinds of tests including the following methods to identify the embryonic stem cells.

- Growing and subculturing the stem cells for many months. This ensures that the cells are capable of long-term growth and self-renewal. Scientists inspect the cultures through a microscope to see that the cells look healthy and remain undifferentiated.

- Using specific techniques to determine the presence of transcription factors that are typically produced by undifferentiated cells. Two of the most important transcription factors are Nanog and Oct4. Transcription factors help turn genes on and off at the right time, which is an important part of the processes of cell differentiation and embryonic development. In this case, both Oct4 and Nanog are associated with maintaining the stem cells in an undifferentiated state, capable of self-renewal.

- Using specific techniques to determine the presence of particular cell-surface markers that are typically produced by undifferentiated cells.

- Examining the chromosomes under a microscope. This is a method to assess whether the chromosomes are damaged or if the number of chromosomes has changed.

- Determining whether the cells can be re-grown, or subcultured, after freezing, thawing, and re-plating.

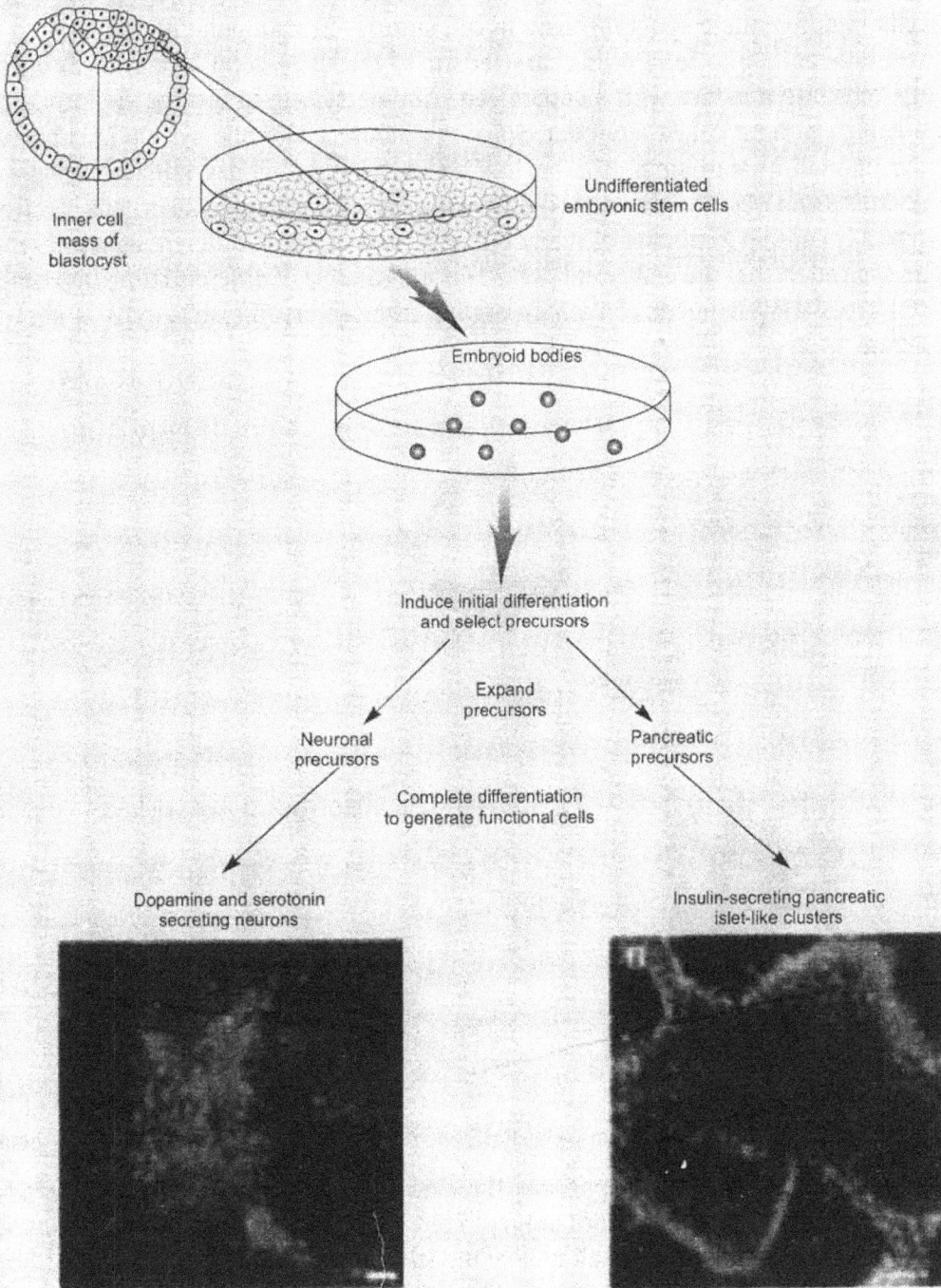

Figure 14.1 Differentiation of mouse embryonic stem cells. (Courtesy: Stem Cell Rejuvenation Center, Phoenix, Arizona, USA).

Testing whether the human embryonic stem cells are pluripotent by 1) allowing the cells to differentiate spontaneously in cell culture; 2) manipulating the cells so they will differentiate to form cells characteristic of the three germ layers; or 3) injecting the cells into a mouse with a suppressed immune system to test for the formation of a benign tumour called teratoma. Since the mouse's immune system is suppressed, the injected human stem cells are not rejected by the mouse immune system and scientists can observe growth and differentiation of the human stem cells. Teratomas typically contain a mixture of many differentiated or partly differentiated cell types— an indication that the embryonic stem cells are capable of differentiating into multiple cell types. Development of various organs from embryonic stem cells is shown in Figure 14.2.

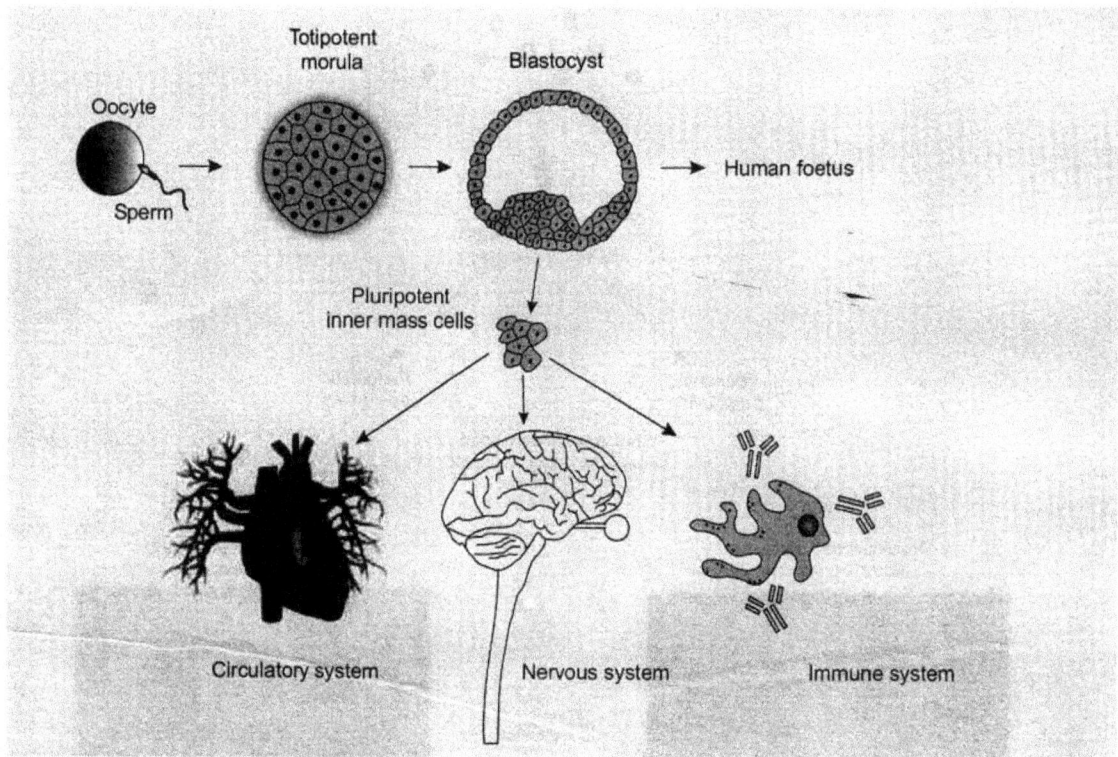

Figure 14.2 Pluripotent embryonic stem cells (Reference: http://en.wikipedia.org/wiki/Cell_potency)

In January of 2009, the FDA approved the first human clinical trial of GRNOPC1, an ES cell-based therapy for spinal cord injury. This study was led by Dr. Hans Keirstead (Keirstead *et al.*, 2005). Based on small animal models, researchers hypothesized that human ES cell-derived oligodendrocyte progenitor cells injected into the damaged spinal cords of patients before the formation of scar tissue will lead to remyelination and the eventual restoration of motor function.

ADULT STEM CELLS

There are also known as somatic (from Greek, "of the body") stem cells and germ line (giving rise to gametes) stem cells, that can be found in children, as well as adults. An adult stem cell is thought to be an undifferentiated cell, found among differentiated cells in a tissue or organ that can renew itself and can differentiate to yield some or all of the major specialized cell types of the tissue or organ. These stem cells have been found in tissues such as the brain, bone marrow, blood, blood vessels, skeletal muscles, skin and the liver. They remain in a quiescent or non-dividing state for years until activated by disease or tissue injury. Pluripotent adult stem cells are rare and generally small in number but can be found in a number of tissues including umbilical cord blood. In mice, plutipotent stem cells are directly generated from adult fibroblast cultures. Most adult stem cells are lineage-restricted (multipotent) and are generally referred to by their tissue of origin (mesenchymal stem cell, adipose-derived stem cell, endothelial stem cell, dental pulp stem cell, etc.). The primary roles of adult stem cells in a living organism are to maintain and repair the tissue in which they are found. Adult stem cell treatments have been successfully used for many years to treat leukaemia and related bone/blood cancers through bone marrow transplants. Adult stem cells are also used in veterinary medicine to treat tendon and ligament injuries in horses. The use of adult stem cells in research and therapy is not as controversial as embryonic stem cells, because the production of adult stem cells does not require the destruction of an embryo. Additionally, because in some instances adult stem cells can be obtained from the intended recipient (an autograft), the risk of rejection is essentially non-existent in these situations. An extremely rich source for adult mesenchymal stem cells is the developing tooth bud of the mandibular third molar. While considered multipotent, they may prove to be pluripotent. The stem cells eventually form enamel (ectoderm), dentin, periodontal ligament, blood vessels, dental pulp, nervous tissues, including a minimum of 29 different unique end organs. These stem cells have been shown to be capable of producing hepatocytes. Differentiation of adult stem cells is shown in Figure 14.3.

Researchers have discovered that the bone marrow contains at least two kinds of stem cells. One population, called haematopoietic stem cells, forms all the types of blood cells in the body. A second population, called bone marrow stromal stem cells (also called mesenchymal stem cells, or skeletal stem cells by some) were discovered a few years later. These non-haematopoietic stem cells make up a small proportion of the stromal cell population in the bone marrow, and can generate bone, cartilage, fat, cells that support the formation of blood, and fibrous connective tissue.

Adult stem cells have been identified in many organs and tissues, including brain, bone marrow, peripheral blood, blood vessels, skeletal muscle, skin, teeth, heart, gut, liver, ovarian epithelium and testis. They are thought to reside in a specific area of each tissue. In many tissues, current evidence suggests that some types of stem cells are pericytes, cells that

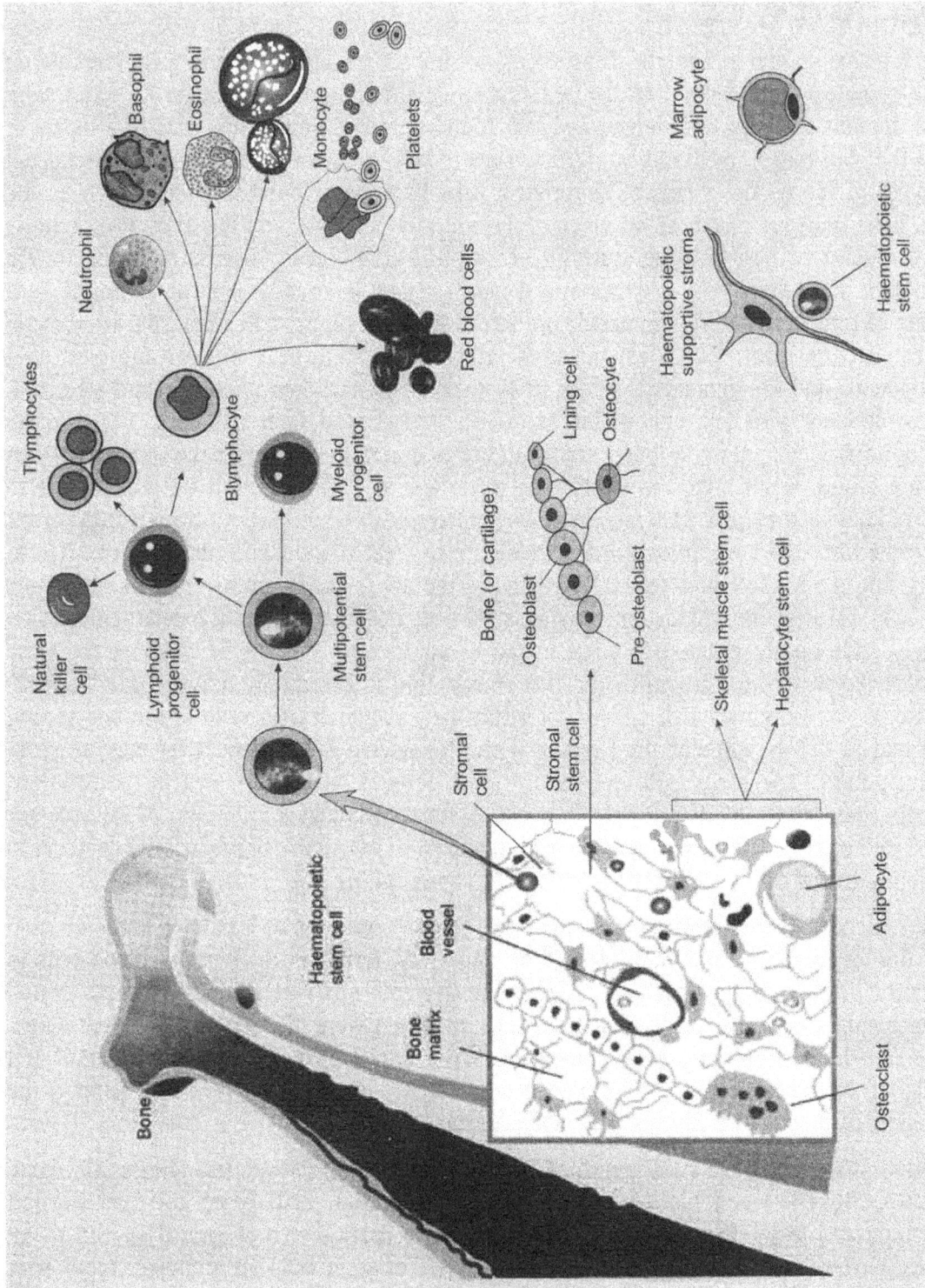

Figure 14.3 Differentiation of adult stem cells (Courtesy: Stem Cell Rejuvenation Center, Phoenix, Arizona, USA).

compose the outermost layer of small blood vessels. Stem cells may remain quiescent (non-dividing) for long periods of time until they are activated by a normal need for more cells to maintain tissues, or by disease or tissue injury. Scientists often use one or more of the following methods to identify adult stem cells: i) label the cells in a living tissue with molecular markers and then determine the specialized cell types they generate; ii) remove the cells from a living animal, label them in cell culture and transplant them back into another animal to determine whether the cells replace their tissue of origin. Importantly, it must be demonstrated that a single adult stem cell can generate a line of genetically identical cells that then gives rise to all the appropriate differentiated cell types of the tissue. Scientists have reported that adult stem cells occur in many tissues and that they enter normal differentiation pathways to form the specialized cell types of the tissue in which they reside.

Normal Differentiation Pathways of Adult Stem Cells

In a living animal, adult stem cells are available to divide. When needed they can give rise to mature cell types that have characteristic shapes and specialized structures and functions of a particular tissue. The following are examples of differentiation pathways of adult stem cells that have been demonstrated *in vitro* or *in vivo*.

- Haematopoietic stem cells give rise to all the types of blood cells: red blood cells, B lymphocytes, T lymphocytes, natural killer cells, neutrophils, basophils, eosinophils, monocytes and macrophages.

- Mesenchymal stem cells give rise to a variety of cell types: bone cells (osteocytes), cartilage cells (chondrocytes), fat cells (adipocytes) and other kinds of connective tissue cells such as those in tendons.

- Neural stem cells in the brain give rise to its three major cell types: nerve cells (neurons), and two categories of non-neuronal cells—astrocytes and oligodendrocytes.

- Epithelial stem cells in the lining of the digestive tract occur in deep crypts and give rise to several cell types: absorptive cells, goblet cells, paneth cells and enteroendocrine cells.

- Skin stem cells occur in the basal layer of the epidermis and at the base of hair follicles. The epidermal stem cells give rise to keratinocytes, which migrate to the surface of the skin and form a protective layer. The follicular stem cells can give rise to both the hair follicle and to the epidermis.

Transdifferentiation

A number of experiments have reported that certain adult stem cell types can differentiate into cell types seen in organs or tissues other than those expected from the cells' predicted lineage (i.e., brain stem cells that differentiate into blood cells or blood-forming cells that differentiate into cardiac muscle cells, and so forth). This reported phenomenon is called transdifferentiation. Although isolated instances of transdifferentiation have been observed in some vertebrate species, whether this phenomenon actually occurs in humans is under debate by the scientific community. Instead of transdifferentiation, the observed instances may involve fusion of a donor cell with a recipient cell. In a variation of transdifferentiation experiments, scientists have recently demonstrated that certain cell types can be "reprogrammed" into other cell types *in vivo* using a well-controlled process of genetic modification. For example, one recent experiment shows how pancreatic beta cells, the insulin-producing cells that are lost or damaged in diabetes, could possibly be created by reprogramming other pancreatic cells. By "re-starting" expression of three critical beta-cell genes in differentiated adult pancreatic exocrine cells, researchers were able to create beta cell-like cells that can secrete insulin. The reprogrammed cells were similar to beta cells in appearance, size, and shape; expressed genes characteristic of beta cells; and were able to partially restore blood sugar regulation in mice whose own beta cells had been chemically destroyed. This method of reprogramming adult cells may be used as a model for directly reprogramming other adult cell types.

It is now possible to reprogram adult somatic cells to become like embryonic stem cells (induced pluripotent stem cells, iPSCs) through the introduction of embryonic genes.

SIMILARITIES AND DIFFERENCES BETWEEN EMBRYONIC AND ADULT STEM CELLS

Human embryonic (ESC) and adult stem cells (ASC) each have advantages and disadvantages regarding potential use for cell-based regenerative therapies. One major difference between adult and embryonic stem cells is their different abilities in the number and type of differentiated cell types they can become. ESC can become all cell types of the body because they are pluripotent. ASC are thought to be limited to differentiating into different cell types of their tissue of origin. ESC can be grown relatively easily in culture. ASC are rare in mature tissues, so isolating these cells from an adult tissue is challenging, and methods to expand their numbers in cell culture have not yet been worked out. This is an important distinction, as large numbers of cells are needed for stem cell replacement therapies. Scientists believe that tissues derived from ESC and ASC may differ in the likelihood of being rejected after transplantation. ASC, and tissues derived from them, are currently believed to be less likely to initiate rejection after transplantation. This is because a patient's own cells could be expanded in culture, coaxed into assuming a specific cell type (differentiation), and then re-introduced into the patient. The use of ASC and tissues derived from the patient's own ASCs would mean that the cells are likely to be rejected by the immune system. This represents a significant advantage, as immune rejection can be circumvented only by continuous administration of immunosuppressive drugs, and the drugs themselves may cause deleterious side effects.

AMNIOTIC STEM CELLS

Multipotent stem cells are also found in amniotic fluid. These stem cells are very active, expand extensively without feeders and are not tumorigenic. Amniotic stem cells are multipotent and can differentiate into cells of adipogenic, osteogenic, myogenic, endothelial, hepatic and also neuronal lines. From an ethical point of view, stem cells from amniotic fluid can solve a lot of problems, because it is possible to catch amniotic stem cells without destroying embryos. It is possible to collect amniotic stem cells for donors or for autologous use: the first amniotic stem cells bank opened in 2009 in Medford, MA, by Biocell Center Corporation and collaborates with various hospitals and universities all over the world.

FOETAL STEM CELLS

Foetal stem cells are self-renewing cells located in various types of foetal tissue, including umbilical cord blood, umbilical cord matrix, foetal blood and the amniotic membrane. Mesenchymal stem cells (MSC) and haematopoietic stem cells are two of the more accessible stem cell populations amongst the foetal stem cell populations. Foetal MSCs can be found in the peripheral blood, bone marrow and liver of the first trimester foetus (Chong, MS and Chan, J., 2010). There are many advantages in using foetal stem cells, which has led scientists to explore these cell types for regenerative therapy. First, the foetal stem cells have shorter doubling times than adult stem cells. They demonstrate greater telomere lengths and their plasticity is superior to that of adult stem cells. Thus, their self-renewal potential may be greater than adult cells and they may possess greater expansion and growth potential without becoming senescent. Another benefit of using foetal stem cells is that umbilical cord blood, which is a major source of foetal stem cells, can be rapidly obtained in emergent situations as "off-the-shelf" stem cell therapy. In addition to these advantages, foetal stem cells appear to be more immunologically naïve than adult stem cells, allowing enhanced transplantation efficiency. They express low levels of MHC class I and nearly undetectable levels of MHC class II, and mismatch between donor foetal stem cells and host tissue is better tolerated than for adult stem cells (Reimann, *et al.*, 2009). The use of foetal stem cells for research and therapeutic purposes carry many ethical questions. For example, foetal stem cells arise from the fusion of sperm and oocyte, and they are frequently obtained from terminated pregnancies or after *in vitro* fertilization, therefore consent cannot be reliably obtained for these foetuses.

INDUCED PLURIPOTENT STEM CELLS

Induced pluripotent stem cells (iPSCs) are adult cells that have been genetically reprogrammed to an embryonic stem cell-like state by being forced to express genes and factors important for maintaining the defining properties of ESCs. Using genetic reprogramming with protein transcription factors, pluripotent stem cells equivalent to embryonic stem cells have been derived from human adult skin tissue. Frozen blood samples can be used as a source of induced pluripotent stem cells, opening a new avenue for obtaining the valued cells. Mouse iPSCs demonstrate important characteristics of pluripotent stem cells, including expressing stem cell markers, forming tumours containing cells from all three germ layers, and being able to contribute to many different tissues when injected into mouse embryos at a very early stage in development. Human iPSCs also express stem cell markers and are capable of generating cells characteristic of all three germ layers. In 2007, Dr. James Thomson of the University of Wisconsin led a team of scientists who were able to derive human-induced pluripotent stem cells through somatic cell nuclear cell transfer and induction with four transcription factors, OCT4, SOX2, NANOG and LIN2822. These cells were demonstrated to be pluripotent, self-renewing and possess normal karyotype and telomerase activity (Yu, J. *et al.*, 2007).

iPSCs are already useful tools for drug development and modelling of diseases, and scientists hope to use them in transplantation medicine. Viruses are currently used to introduce the reprogramming factors into adult cells, and this process must be carefully controlled and tested before the technique can lead to useful treatments for humans. In animal studies, the virus used to introduce the stem cell factors sometimes causes cancers. Researchers are currently investigating non-viral delivery strategies. In any case, this breakthrough discovery has created a powerful new way to "de-differentiate" cells whose developmental fates had been previously assumed to be determined. In addition, tissues derived from iPSCs will be a nearly identical match to the cell donor and thus probably avoid rejection by the immune system.

POTENTIAL USES OF HUMAN STEM CELLS

There are many ways in which human stem cells can be used in research and the clinic. Studies of human ESCs will yield information about the complex events that occur during human development. A primary goal of this work is to identify how undifferentiated stem cells become the differentiated cells that form the tissues and organs. Scientists know that turning genes on and off is central to this process. Some of the most serious medical conditions, such as cancer and birth defects, are due to abnormal cell division and differentiation. A more complete understanding of the genetic and molecular controls of these processes may yield information about how such diseases arise and suggest new strategies for therapy. Predictably, controlling cell proliferation and differentiation requires additional basic research on the molecular and genetic signals that regulate cell division and specialization.

HSCs could also be used to test new drugs. For example, new medications could be tested for safety on differentiated cells generated from human pluripotent cell lines. Other kinds of cell lines are already used in this way. Cancer cell lines, for example, are used to screen potential anti-tumour drugs. The availability of pluripotent stem cells would allow drug testing in wider range of cell types. However, to screen drugs effectively, the conditions must be identical when comparing different drugs. Therefore, scientists will have to be able to precisely control the differentiation of stem cells into the specific cell type on which drugs will be tested. Perhaps the most important potential application of human stem cells is the generation of cells and tissues that could be used for cell-based therapies. Today, donated organs and tissues are often used to replace ailing or destroyed tissue, but the need for transplantable tissues and organs far outweighs the available supply. Stem cells, directed to differentiate into specific cell types, offer the possibility of a renewable source of replacement cells and tissues to treat diseases including Alzheimer's disease, spinal cord injury, stroke, burns, heart disease, diabetes, osteoarthritis and rheumatoid arthritis. For example, it may become possible to generate healthy heart muscle cells in the laboratory and then transplant those cells into patients with chronic heart disease. Preliminary research in mice and other animals indicates that bone marrow stromal cells, transplanted into a damaged heart can have beneficial effects. Whether these cells can generate heart muscle cells or stimulate the growth of new blood vessels that repopulate the heart tissue, or help via some other mechanism is actively under investigation. For example, injected cells may accomplish repair by secreting growth factors, rather than actually incorporating into the heart. Promising results from animal studies have served as the basis for a small number of exploratory studies in humans. Other recent studies in cell culture systems indicate that it may be possible to direct the differentiation of embryonic stem cells or adult bone marrow cells into heart muscle cells. In people who suffer from type 1 diabetes, the cells of the pancreas that normally produce insulin are destroyed by the patient's own immune system. New studies indicate that it may be possible to direct the differentiation of human embryonic stem cells in cell culture to form insulin-producing cells that eventually could be used in transplantation therapy for persons with diabetes.

ORGAN AND TISSUE REGENERATION

Tissue regeneration is probably the most important possible application of stem cell research. Currently, organs must be donated and transplanted, but the demand for organs far exceeds supply. Stem cells could potentially be used to grow a particular type of tissue or organ if directed to differentiate in a certain way. Stem cells that lie just beneath the skin, for example, have been used to engineer new skin tissue that can be grafted on to burn victims.

TREATMENTS

Bone marrow transplantation As of 2009, bone marrow transplantation is the only established use of stem cells. Medical researchers believe that stem cell therapy has the potential to dramatically change the treatment of human disease. A number of adult stem cell therapies already exist, particularly bone marrow transplants that are used to treat leukaemia. In the future, medical researchers anticipate being able to use technologies derived from stem cell research to treat a wider variety of diseases including cancer, Parkinson's disease, spinal cord injuries, amylotrophic lateral sclerosis, multiple sclerosis and muscle damage, amongst a number of other impairments and conditions. One concern of treatment is the possible risk that transplanted stem cells could form tumours and have the possibility of becoming cancerous if cell division continues uncontrollably. Supporters of embryonic stem cell research argue that research should be pursued because the resultant treatments could have significant medical potential. It is also noted that excess embryos created for *in vitro* fertilization could be donated with consent and used for the research.

Brain disease treatment　Clinical trials based on stem replacement therapy are confined to patients with Parkinson's disease (PD), Huntington's disease (HD), amylotrophic lateral sclerosis (ALS), and some lysosomal storage disorders. Though results were varying, in all cases, stem cells were able to integrate into the host brain tissue, with positive effects for the recipients and, in a few cases, stem cells were able to restore the degenerated tissue. Additionally, replacement cells and tissues may be used to treat brain disease such as Parkinson's and Alzheimer's by replenishing damaged tissue, bringing back the specialized brain cells that keep unneeded muscles from moving. Embryonic stem cells have recently been directed to differentiate into these types of cells, and so treatments are promising.

Cell deficiency therapy　Healthy heart cells developed in a laboratory may one day be transplanted into patients with heart disease, repopulating the heart with healthy tissue. Similarly, people with type I diabetes may receive pancreatic cells to replace the insulin-producing cells that have been lost or destroyed by the patient's own immune system. The only current therapy is a pancreatic transplant, and it is unlikely to occur due to a small supply of pancreases available for transplant.

Blood disease treatments　Adult haematopoietic stem cells found in blood and bone marrow have been used for years to treat diseases such as leukemia, sickle cell anaemia and other immunodeficiencies. These cells are capable of producing all blood cell types—from red blood cells that carry oxygen to white blood cells that fight disease. Difficulties arise in the extraction of these cells through the use of invasive bone marrow transplants. However, haematopoietic stem cells have also been found in the umbilical cord and placenta. This has led some scientists to call for an umbilical cord blood bank to make these powerful cells more easily obtaintable and to decrease the chances of a body's rejecting therapy.

General scientific discovery　Stem cell research is also useful for learning about human development. Undifferentiated cell eventually differentiate partly because a particular gene is turned on or off. Stem cell researchers may help to clarify the role that genes play in determining what genetic traits or mutations we receive. Cancer and other birth defects are also affected by abnormal cell division and differentiation. New therapies for diseases may be developed if we better understand how these agents attack the human body. Another reason why stem cell research is being pursued is to develop new drugs. Scientists could measure a drug's effect on healthy, normal tissues by testing the drug on tissue grown from stem cells rather than testing the drug on human volunteers.

STEM CELL CONTROVERSY

The debates surrounding stem cell research primarily are driven by methods concerning embryonic stem cell research. It was only in 1998 that researchers from the University of Wisconsin-Madison extracted the first human embryonic stem cells that were able to be kept alive in the laboratory. The main critique of this research is that it required the destruction of a human blastocyst. That is, a fertilized egg was not given the chance to develop into a fully developed human.

Table 14.1 Ongoing research in stem cells around the world

Diseases /conditions	Place where research is conducted
Amylotropic lateral sclerosis and spinal cord injury	Stanford Medical School, USA
Blindness	University of California, Irvine, USA Columbia University Medical Center, USA Hadassah-Hebrew University Medical Center, Jerusalem University College London, Great Britain
Blood supply	Wellcome Trust in Great Britain
Brain damage	Stanford University, USA
Cancer	University of Connecticut Stem Cell Institute, USA National Cancer Institute, USA University of Minnesota, USA
Cartilage damage	University of California Davis, USA
Diabetes	Harvard Stem Cell Institute
Hearing loss	Stanford University, USA
Heart disease and defects	Harvard University, USA University of Washington, USA Harvard Stem Cell Institute, USA
Lung damage	Free University of Brussels, Belgium
Memory loss	University of California, Irvine, USA
Stroke	Stanford University, USA
Tissue damage	Institute for Stem Cell Therapy and Exploration of Monogenic Diseases in France
Neonatology and family medicine	Loma Linda University

REVIEW QUESTIONS

1. Write in detail the properties of stem cells.

2. What are the potential uses of human stem cells?

3. Write short notes on the following:

    i. Embryonic stem cells

    ii. Adult stem cells

    iii. Foetal stem cells

GLOSSARY

Adjuvant A substance that enhances the body's immune response to an antigen. It helps and enhances the pharmacological effect of a drug or increases the ability of an antigen to stimulate the immune system.

Adventitious Developing from unusual points of origin, such as embryos from sources other than zygotes. This term can also be used to describe agents which contaminate cell cultures.

Anchorage-dependent cells or cultures Cells,or cultures derived from them, which will grow, survive, or maintain function only when attached to a surface such as glass or plastic. The use of this term does not imply that the cells are normal or that they are or are not neoplastically transformed.

Aneuploid The situation which exists when the nucleus of a cell does not contain an exact multiple of the haploid number of chromosomes, but has one or more chromosomes present in greater or lesser number than the rest. The chromosomes may or may not show rearrangements.

Apoptosis Cell death by a biologically controlled intracellular process which involves cleavage of DNA and nuclear fragmentation.

Asepsis Without infection or contaminating microorganisms.

Aseptic technique Procedures used to prevent the introduction of fungi, bacteria, viruses, mycoplasma or other microorganisms into cell, tissue and organ culture. Although these procedures are used to prevent microbial contamination of cultures, they also prevent cross contamination of cell cultures. These procedures may or may not exclude the introduction of infectious molecules.

Attachment efficiency The percentage of cells plated (seeded, inoculated) which attach to the surface of the culture vessel within a specified period of time. The conditions under which such a determination is made should always be stated.

Autocrine cell In animals, a cell which produces hormones, growth factors or other signalling substances for which it also expresses the corresponding receptors.

Axenic culture A culture without foreign or undesired life forms. An axenic culture may include the purposeful co-cultivation of different types of cells, tissues or organisms.

Balance salt solution An isotonic solution of inorganic salts present in correct physiological concentrations which contains glucose which is usually free from other organic nutrients.

Bioreactor Culture vessel for large-scale production of cells either anchored to the substrate or propagated in suspension. This can be used for small-scale three-dimensional culture of constructs for tissue engineering.

Biostat   Culture vessel in which physical, physiological and physiochemical conditions, as well as cell concentration are kept constant, usually by perfusion, monitoring and feedback.

Caspase assay   An enzyme that plays a key role in programmed cell death or apoptosis. The caspase 3 assay is based on the hydrolysis of acetyl-Asp-Glu-Val-Asp r-nitroanilidine by caspase 3, resulting in the release of the r-nitroaniline moiety. r-Nitroaniline is detected at 405 nm.

Cell culture   Term used to denote the maintenance or cultivation of cells *in vitro* including the culture of single cells. In cell cultures, the cells are no longer organized into tissues.

Cell density   Number of cells per cm^2 of substrate.

Cell generation time   The interval between consecutive divisions of a cell. This interval can best be determined, at present, with the aid of cinephotomicrography. *(This term is not synonymous with "population doubling time".)*

Cell hydridization   The fusion of two or more dissimilar cells leading to the formation of a synkaryon.

Cell line   A cell line arises from a primary culture at the time of the first successful subculture. The term "cell line" implies that cultures from it consist of lineages of cells originally present in the primary culture. The terms *finite* or *continuous* are used as prefixes if the status of the culture is known. If not, the term *line* will suffice. The term *"continuous line" replaces the term "established line"*. In any published description of a culture, one must make every attempt to publish the characterization or history of the culture. If such has already been published, a reference to the original publication must be made. In obtaining a culture from another laboratory, the proper

designation of the culture, as originally named and described, must be maintained and any deviations in cultivation from the original must be reported in any publication.

Cell strain   A cell strain is derived either from a primary culture or a cell line by the selection or cloning of cells having specific properties or markers. In describing a cell strain, its specific features must be defined. The terms finite or continuous are to be used as prefixes if the status of the culture is known. If not, the term *strain* will suffice. In any published description of a cell strain, one must make every attempt to publish the characterization or history of the strain. If such has already been published, a reference to the original publication must be made. In obtaining a culture from another laboratory, the proper designation of the culture, as originally named and described, must be maintained and any deviations in cultivation from the original must be reported in any publication.

Chemically defined medium   A nutritive solution for culturing cells in which each component is specifiable, and ideally is of known chemical structure.

Clone   In animal cell culture terminology a population of cells derived from a single cell by mitoses. A clone is not necessarily homogeneous and therefore, the terms *clone* and *cloned* do not indicate homogeneity in a cell population, genetic or otherwise.

Cloning efficiency   The percentage of cells plated (seeded, inoculated) that form a clone. One must be certain that the colonies formed arose from single cells.

Colony forming efficiency   The percentage of cells plated (seeded, inoculated) that form a colony.

Complementation   The ability of two different genetic defects to compensate for one another.

Contact inhibition of locomotion A phenomenon characterizing certain cells in which two cells meet, locomotory activity diminishes, and the forward motion of one cell over the surface of the other is stopped.

Continuous cell culture A culture which is apparently capable of an unlimited number of population doublings; often referred to as an immortal cell culture. Such cells may or may not express the characteristics of *in vitro* neoplastic or malignant transformation.

Crisis A stage in the *in vitro* transformation of cells characterized by reduced proliferation of the culture, abnormal mitotic figures, detachment of cells from the culture substrate, and the formation of multinucleated or giant cells. During this massive cultural degeneration, a small number of colonies usually, but not always, survive and give rise to a culture with an apparent unlimited *in vitro* lifespan. This process was first described in human cells following infection with an oncogenic virus (SV40).

Cryopreservation Ultra-low-temperature storage of cells, tissues, embryos or seeds. This storage is usually carried out using temperatures below $-100°C$.

Cumulative population doublings (*See* Population doubling level.)

Cybrid The viable cell resulting from the fusion of a cytoplast with a whole cell, thus creating a cytoplasmic hybrid.

Cytoplasmic inheritance Inheritance attributable to extranuclear genes; for example genes in cytoplasmic organelles such as mitochondria or chloroplasts, or in plasmids, etc.

Density-dependent inhibition of growth Mitotic inhibition correlated with increased cell density.

Differentiated Cells that maintain, in culture, all or much of the specialized structure and function typical of the cell type *in vivo*.

Diploid The state of the cell in which all chromosomes, except sex chromosomes, are two in number and are structurally identical with those of the species from which the culture was derived. Where there is a Commission Report available, the experimenter should adhere to the convention for reporting the karyotype of the donor. Commission Reports have been published for mouse, human, and rat. In defining a diploid culture, one should present a graph depicting the chromosome number distribution leading to the modal number determination along with representative karyotypes.

Disinfection The process of destroying pathogenic organisms or rendering them inert.

Electroporation Creation, by means of an electrical current, of transient pores in the plasmalemma usually for the purpose of introducing exogenous material, especially DNA, from the medium.

Embryo culture *In vitro* development or maintenance of isolated mature or immature embryos.

Embryogenesis The process of embryo initiation and development.

Endocrine cell In animals, a cell which produces hormones, growth factors or other signalling substances for which target cells, expressing the corresponding receptors, are located at a distance.

Epigenetic event Any change in a phenotype which does not result from an alteration in DNA sequence. This change may be stable and heritable and includes alteration in DNA methylation, transcriptional activation, translational control and post-translational modifications.

Epigenetic variation Phenotypic variability which has a nongenetic basis.

Epithelial-like Resembling or characteristic of, having the form or appearance of epithelial cells. In order to define a cell as an epithelial cell, it must possess characteristics typical of epithelial cells. Often one can be certain of the histologic origin and/or function of the cells placed into culture and, under these conditions, one can be reasonably confident in designating the cells as epithelial. It is incumbent upon the individual reporting on such cells to use as many parameters as possible in assigning this term to a culture. Until such time as a rigorous definition is possible, it would be most correct to use the term "epithelial-like".

Euploid The situation which exists when the nucleus of a cell contains exact multiples of the haploid number of chromosomes.

Explant Tissue taken from its original site and transferred to an artificial medium for growth or maintenance.

Explant culture The maintenance or growth of an explant in culture.

Feeder layer A layer of cells (usually lethally irradiated for animal cell culture) upon which are cultured a fastidious cell type.

Fermenter Large-scale culture vessel often used for cells in suspension,derived from same term applied to microbiological cultures.

Fibroblast A proliferating precursor cell of the mature differentiated fibrocyte.

Fibroblast-like Resembling or characteristic of, having the form or appearance of fibroblast cells. In order to define a cell as a fibroblast cell, it must possess characteristics typical of fibroblast cells. Often one can be certain of the histologic origin and/or function of the cells placed into culture and, under these conditions, one can be reasonably confident in designating the cells as fibroblast. It is incumbent upon the individual reporting on such cells to use as many parameters as possible

in assigning this term to a culture. Until such time as a rigorous definition is possible, it would be most correct to use the term "fibroblast-like."

Finite cell culture A culture which is capable of only a limited number of population doubling after which the culture ceases proliferation.

Finite cell line A culture that has been propagated by subculture but is capable of only a limited number of cell generations *in vitro* before dying.

Genetic engineering The deliberate, controlled manipulation of the genes in an organism with the intent of making that organism better in some way.

Gene therapy The insertion, alteration, or removal of genes within an individual's cells and biological tissues to treat disease. It is a technique for correcting defective genes that are responsible for disease development.

Growth curve A semi-logarithmic plot of the cell number on a logarithmic scale against time on a linear scale for a proliferating cell culture.

Growth medium The medium used to propagate a particular cell line, usually a basal medium with additives such as serum or growth factors.

HEPA filter High efficiency particulate air filter. A filter capable of screening out particles larger than $0.3\,\mu$m. HEPA filters are used in laminar air flow cabinets (hoods) for sterile transfer work.

Heterokaryon A cell possessing two or more genetically different nuclei in a common cytoplasm, usually derived as a result of cell-to-cell fusion.

Heteroploid The term given to a cell culture when the cells comprising the culture possess nuclei containing chromosome numbers other than the diploid number. This is a term used only to describe a culture and is not used to describe individual cells. Thus, a heteroploid culture would be one which contains aneuploid cells.

Histiotypic The *in vitro* resemblance of cells in culture to a tissue in form or function or both. For example, a suspension of fibroblast–like cells may secrete a glycosaminoglycan–collagen matrix and the result is a structure resembling fibrous connective tissue, which is, therefore, histiotypic. This term is not meant to be used along with the word "culture". Thus, a tissue culture system demonstrating form and function typical of cells *in vivo* would be said to be histiotypic.

Homokaryon A cell possessing two or more genetically identical nuclei in a common cytoplasm, derived as a result of cell-to-cell fusion.

Hybrid cell The term used to describe the mononucleate cell which results from the fusion of two different cells, leading to the formation of a synkaryon.

Hybridoma The cell which results from the fusion of an antibody-producing tumour cell (myeloma) and an antigenically stimulated normal plasma cell. Such cells are constructed because they produce a single antibody directed against the antigen epitope which stimulated the plasma cell. This antibody is referred to as a monoclonal antibody.

Ideogram The arrangement of the chromosome of a cell in order by size and morphology so that the karyotype may be studied and genetically analysed.

Immortalization The attainment by a finite cell culture, whether by perturbation or intrinsically, of the attributes of a continuous cell line. An immortalized cell is not necessarily one which is neoplastically or malignantly transformed.

In vitro **neoplastic transformation** The acquisition, by cultured cells, of the property to form neoplasms, benign or malignant, when inoculated

into animals. Many transformed cell populations which arise *in vitro* intrinsically or through deliberate manipulation by the investigator, produce only benign tumours which show no local invasion or metastasis following animal inoculation. If there is supporting evidence, the term "*in vitro* malignant neoplastic transformation" or " *in vitro* malignant transformation" can be used to indicate that an injected cell line does, indeed, invade or metastasize.

In vitro **senescence** In vertebrate cell cultures, the property attributable to finite cell cultures namely, their inability to grow beyond a finite number of population doublings. Invertebrate cell cultures do not exhibit this property.

In vitro **transformation** A heritable change occurring in cells in culture either intrinsically or from treatment with chemical carcinogens, oncogenic viruses, irradiation, transfection with oncogenes, etc. and leading to the acquisition of altered morphological, antigenic, neoplastic, proliferative or other properties. This expression is distinguished from "*in vitro* neoplastic transformation" in that the alterations occurring in the cell population may not always include the ability of the cells to produce tumours in appropriate hosts. The type of transformation should always be specified in any description.

iPS cells A plutipotent stem cell induced from genetic manipulation and /or epigenetic regulation of gene expression of adult cells.

Karyoplast A cell nucleus, obtained from the cell by enucleation, surrounded by a narrow rim of cytoplasm and a plasma membrane.

Liposome A closed lipid vesicle surrounding an aqueous interior; may be used to encapsulate exogenous materials for ultimate delivery of these into cells by fusion with the cell.

Microcell A cell fragment, containing one to a few chromosomes, which is formed by the enucleation or disruption of a micronucleated cell.

Micronucleated cell A cell which has been mitotically arrested and in which small groups of chromosomes function as foci for the reassembly of the nuclear membrane thus forming micronuclei, the maximum of which could be equal to the total number of chromosomes.

Morphogenesis The evolution of a structure from an undifferentiated to a differentiated state; the process of growth and development of differentiated structures.

Mutant A phenotypic variant resulting from a changed or new gene.

Myeloma A malignant tumour of antibody-producing cells, called plasma cells, that are normally found in the bone marrow.

Organ culture The maintenance or growth of organ primordia or the whole or parts of an organ *in vitro* in a way that may allow differentiation and preservation of the architecture and/or function.

Organized Arranged into definite structures.

Organogenesis The evolution, from dissociated cells, of a structure which shows natural organ form or function or both.

Organotypic Resembling an organ *in vivo* in three-dimensional form or function or both. For example, a rudimentary organ in culture may differentiate in an *organotypic* manner, or a population of dispersed cells may become rearranged into an *organotypic* structure and may also function in an *organotypic* manner. This term is not meant to be used along with the word "culture" but is meant to be used as a descriptive term.

Passage The transfer or transplantation of cell, with or without dilution, from one culture vessel to another. It is understood that any time cells are transferred from one vessel to another, a certain portion of the cells may be lost and, therefore, dilution of cells, whether deliberate or not, may occur. This term is synonymous with the term "subculture".

Passage number The number of times the cells in the culture have been subcultured or passaged. In descriptions of this process, the ratio or dilution of the cells should be stated so that the relative cultural age can be ascertained.

Pathogen-free Free from specific organisms based on specific tests for the designated organisms.

Plating efficiency A term which originally encompasses the terms attachment (seeding) efficiency, cloning efficiency, and colony forming efficiency and which is now better described by using one or more of them in its place as the term "plating" is not sufficiently descriptive of what is taking place.

Population density The number of cells per unit area or volume of a culture vessel. Also the number of cells per unit volume of medium in a suspension culture.

Population doubling level The total number of population doubling of a cell line or strain since its initiation *in vitro*. A formula to use for the calculation of "population doublings" in a single passage is:

Number of population doublings $= \log_{10}(N/N_0) \times 3.33$

where, $N =$ number of cells in the growth vessel at the end of a period of growth, $N_0 =$ number of cells plated in the growth vessel. It is best to use the number of viable cells or number of attached cells for this determination. Population doubling level is synonymous with "cumulative population doublings."

Population doubling time The interval, calculated during the logarithmic phase of growth in which, for example, 1.0×10^6 cells increase to 2.0×10^6 cells. This term is not synonymous with "cumulative population doublings".

Primary culture A culture started from cells, tissues or organs taken directly from organisms. A primary culture may be regarded as such until it is successfully subcultured for the first time. It then becomes a "cell line".

Protoplast fusion Technique in which protoplasts are fused into a single cell.

Pseudodiploid This describes the condition where the number of chromosomes in a cell is diploid but, as a result of chromosomal rearrangements, the karyotype is abnormal and linkage relationships may be disrupted.

Recon The viable cell reconstructed by the fusion of a karyoplast with a cytoplast.

Regeneration A morphogenetic response to a stimulus that results in the production of cells with normal and specific characteristics.

Saturation density The maximum cell number attainable, under specified culture conditions, in a culture vessel. This term is usually expressed as the number of cells per square centimetre in a monolayer culture or the number of cells per cubic centimetre in a suspension culture.

Somatic cell genetics The study of genetic phenomena of somatic cells. The cells under study are most often cells grown in culture.

Somatic cell hybridization The *in vitro* fusion of animal cells derived from somatic cells which differ genetically.

Somatic embryogenesis The process of embryo initiation and development from vegetative or nongametic cells.

Sparging The bubbling of gas through the culture and is an efficient means of effecting oxygen transfer.

Spheroid A three-dimensional cluster of cells formed by reaggregation of cells in suspension, usually over a non-adhesive substrate such as agar or agarose.

Split ratio The divisor of the dilution ratio of a cell culture at subculture. (e.g., if one flask content is divided into four flasks then split ratio is 1 : 4)

Substrain A strain that can be derived from a strain by isolation of a single cell or groups of cells having properties or markers not shared by all cells of the parent strain.

Suspension culture A type of culture in which cells, or aggregates of cells, multiply while suspended in liquid medium.

Synkaryon A hybrid cell which results from the fusion of the nuclei it carries.

Three-dimensional culture Culture of cells as aggregates or in a matrix or scaffold such that the cells are in three-dimensional array of standard monolayer culture.

Tissue culture The maintenance or growth of tissues, *in vitro*, in a way that may allow differentiation and preservation of their architecture and/or function.

Totipotency A cell characteristic in which the potential for forming all the cell types in the adult organism is retained.

Transduction Transfer of genetic material or characteristics from one bacterial cell to another by the incorporation of bacterial DNA into a bacteriophage.

Transfection The transfer of naked, foreign DNA into cells in culture for the purpose of genomic integration. The definition as stated here is that which is in use to describe the general transfer of DNA irrespective of its source.

Transformation The introduction and stable genomic integration of foreign DNA into an animal cell by any means, resulting in a genetic modification.

TUNEL assay A common method for detecting DNA fragmentation that results from apoptotic signalling cascades. The assay relies on the presence of nicks in the DNA which can be identified by terminal deoxynucleotidyl transferase, an enzyme that will catalyse the addition of dUTPs that are secondarily labelled with a marker.

REFERENCES

Barnes, D. and Sato, G. (1980). "Methods for growth of cultured cells in serum-free medium." *Anal. Biochem.* 102 : 255–270.

Barnes, W.D., Sirbasku, D.A. and Sato, G.H. (Eds.). (1984a). *Cell Culture Methods for Molecular and Cell Biology, Vol. 1. Methods for Preparation of Media, Supplements, and Substrata for Serum-Free Animal Cell Culture.* Alan R. Liss, Inc., New York.

Barnes, W.D., Sirbasku, D.A. and Sato, G.H. (Eds.). (1984b). *Cell Culture Methods for Molecular and Cell Biology, Vol. 2. Methods for Serum-Free Culture of Cells of the Endocrine System.* Alan R. Liss, Inc., New York.

Beddington, R. (1992). "Transgenic mutagenesis in the mouse." *Trends Genetics.* 8:10.

Chong, M.S. and Chan, J. (2010). "Lentiviral vector transduction of fetal mesenchymal stem cells." *Methods Mol. Biol.* 614: 135–147.

Coons, A.H. and Kaplan, M.M. (1950). "Localization of antigen in tissue cells. II. Improvements in a method for the detection of antigen by means of fluorescent antibody." *J. Exp. Med.* 91 : 1–13.

Eunan McGlinchey. (2007). "Animal Cell Culture Scale-up." *Encyclopedia of Life Sciences.* John Wiley & Sons, Ltd.

Freshney, R.I., Pragnell, I.B. and Freshney, M.G. (1994). *Culture of Haemopoietic Cells.* Wiley-Liss, New York.

Freshney, R.I. (1985). "Induction of differentiation in neoplastic cells." *Anticancer Res.* 5:111–130.

Freshney, R.I. (Ed.). (1992). *Culture of Epithelial Cells.* Wiley-Liss, New York.

Freshney, R.I. (Ed.). (2007). *Culture of Animal Cells.* Wiley-Blackwell, New York.

Griffiths, B. (1992) "Scaling-up of animal cell cultures." In: Freshney, R.I. (Ed.). *Animal Cell Culture: A Practical Approach.* IRL Press, Oxford. pp. 47–93.

Ham, R.G. and McKeehan, W.L. (1979) "Media and growth requirements." In: Jakoby, W.B. and Pastan, I.H. (Eds.). *Methods in Enzymology, Vol. 58. Cell Culture*. Academic Press, New York, pp. 44–93.

Hay, R.J. (1992). "Cell line preservation and characterization." In: Freshney, R.I. (Ed.). *Culture of Animal Cells, A Practical Approach*. IRL Press, Oxford University Press, Oxford. pp. 95–148.

Johnstone, A. and Thorpe, R. (1987). *Immunocytochemistry in Practice*. Blackwell Scientific Publications, Oxford.

Keirstead, H.S. *et al.* (2005). "Human embryonic stem cell-derived oligodendrocyte progenitor cell transplants remyelinate and restore locomotion after spinal cord injury." *J. Neurosci*. 25: 4694–4705.

Kruse, P. and Patterson, M.K. (1973). *Tissue Culture Techniques and Applications*. Academic Press, New York.

Kruse, P.F., Jr. and Miedema, E. (1965). "Production and characterization of multiple-layered populations of animal cells." *J. Cell Biol*. 27: 273.

Kruse, P.F., Jr., Keen, L.N. and Whittle, W.L. (1970). "Some distinctive characteristics of high density perfusion cultures of diverse cell types." *In Vitro*. 6:75–78.

Kuchler, R.J. (Ed.). (1974). *Animal Cell Culture and Virology*. Dowden, Hutchinson & Ross, Stroudsburg, PA.

Kuchler, R.J. (1977) *Biochemical Methods in Cell Culture and Virology*. Academic Press, New York.

Lasfargues, E.Y. (1973). "Human mammary tumors." In: Kruse, P. and Patterson, M.K. (Eds.). *Tissue Culture Methods and Applications*. Academic Press, New York, pp. 45–50.

Lubiniecki, A.S. (Ed.). (1990). *Large-scale Mammalian Cell Culture Technology, Vol. 10. Bioprocess Technology*. Marcel Dekker, New York.

Reimann, V., Creutzig, U. and Kogler, G. (2009). "Stem cells derived from cord blood in transplantation and regenerative medicine." *Dtsch Arztebl Int*. 106: 831–836.

Safe Working and the Prevention of Infection in Clinical Laboratories (1991). Health and Safety Commission, HMSO Publications, London, SW8 5DT, England.

Sambrook, J., Fritsch, E.F. and Maniatis, T. (1989). *Molecular Cloning. A Laboratory Manual*, 2nd edn. 3 Vols. Cold Spring Harbor Laboratory Press, Cold Spring Harbor, NY.

Triglia, D., Braa, S.S., Yonan, C. and Naughton, G.K. (1991). "Cytotoxicity testing using neutral red and MTT assays on a three-dimensional human skin substrate." *Toxic In Vitro*. 5:573–578.

Watt, F. (1991). Annual Meeting of European Tissue Culture Society, Krackow, Poland.

Yu, J. *et al.* (2007). "Induced pluripotent stem cell lines derived from human somatic cells." *Science*. 318: 1917–1920.

http://www.stemcellactionnetwork.org/content/recent-advances-human-embryonic-stem-cell-research

INDEX

B lymphoblast cell line—B95

HeLa cells—Normal (40×)

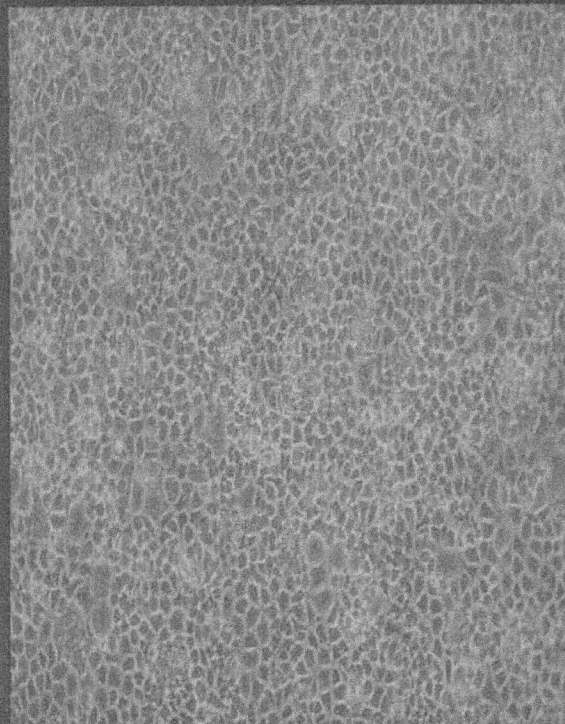
HeLa cells—Infected showing focal rounding (40×)

Human neuroblastoma cell line—Normal (40×)

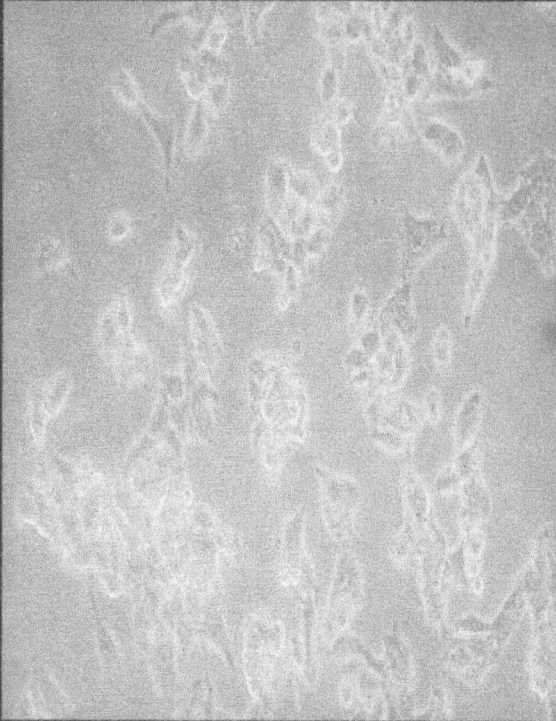
Human neuroblastoma cell line—
Infected showing cell granuation (40×)

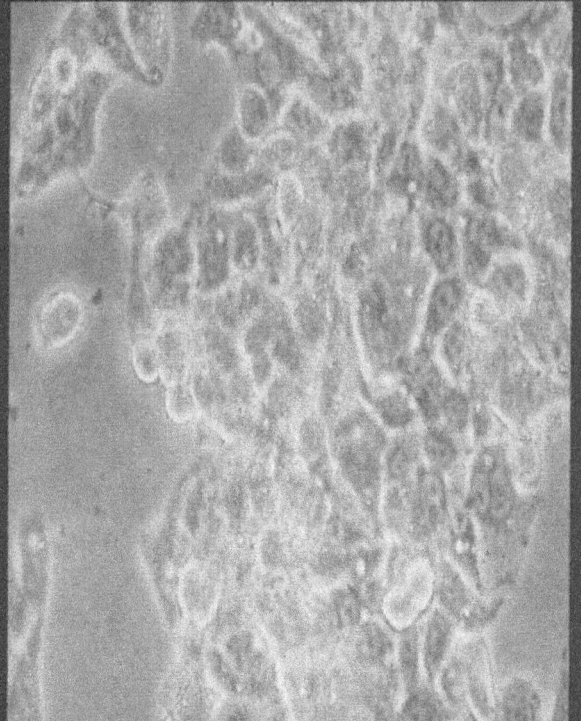
Human neuroblastoma cell line—
Infected showing cell granuation (200×)

THP1 cell line (Human monocytic Lukemia cell line) (20×)

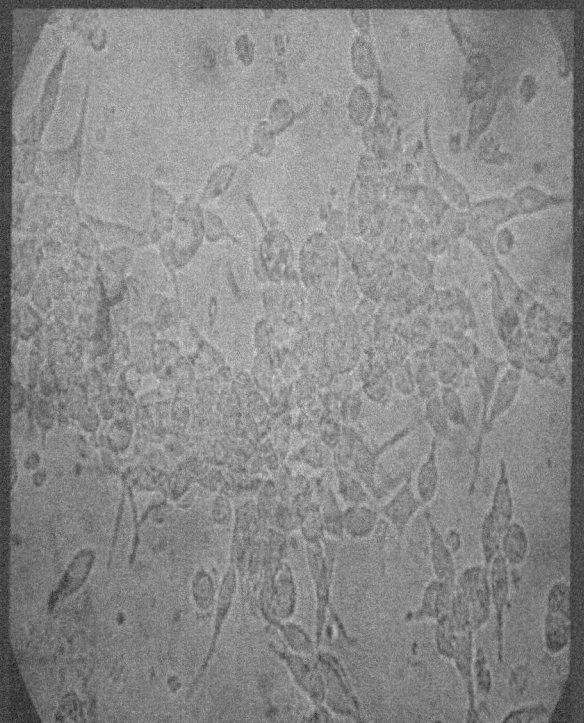
MDBK cell line—Infected (40×)

Mammary cell line HTB—133 showing focal clumping and cell depletion of cells due to the effect of a plant based anticancerous extract.(40×)

Normal MCF-7 human mammary adenocarcinoma cell line 32 hrs (40×)

Effect of the anti-cancerous suspect on MCF-7 cell lines—12 hrs post-infection showing cell rounding (20×)

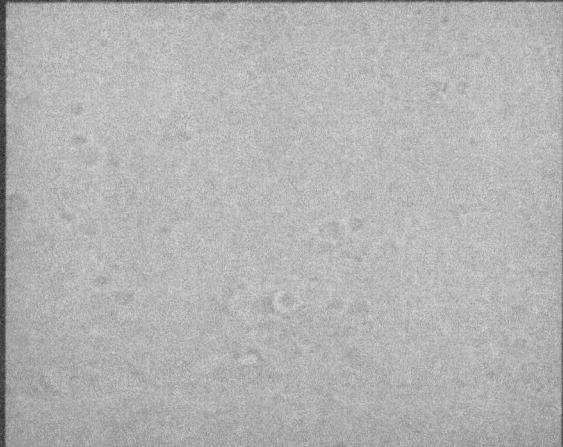
Effect of the anti-cancerous suspect on MCF-7 cell lines—12 hrs post-infection showing cell rounding (40×)

A non malignant transformed mammary cell line— 184A1 showing anticancerous effect of adrug (40X)

Effect of the anti-cancerous suspect on MCF-7 cell lines—24 hrs post-infection showing cell rounding (100×)

Effect of the an anti-cancerous metabolite
on MCF-7 cell lines—32 hrs post-infection
showing cell growth inhibition and prevention (40×)

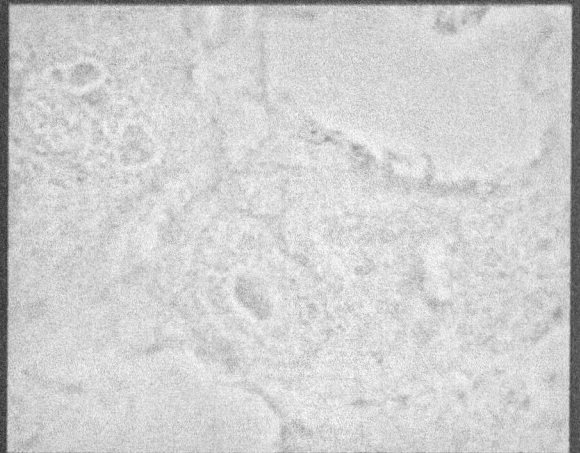
A6 cells cell line derived from the
kidney of a male frog (Xenopus laevis)

MTT assay of anti-cancerous substance in cancerous
cell line—Formation of reactive intermediate (blue crystals)

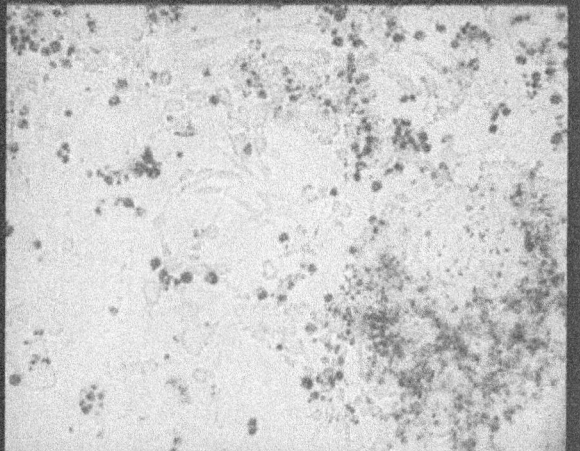
Cell contaminants stained using Hoechst stain (200×)

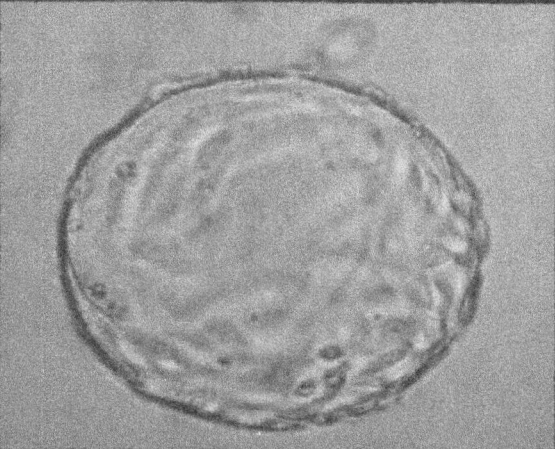
Microcarrier culture showing attachment of
cells on the surface of the carrier beads

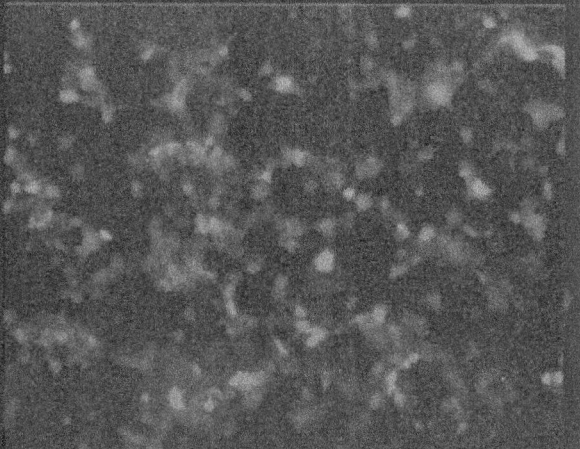
HeLa cells infected with Polio virus showing
fluorescence of viral particles (20×, FITC)

Human tooth root tissue fibroblast (20×)

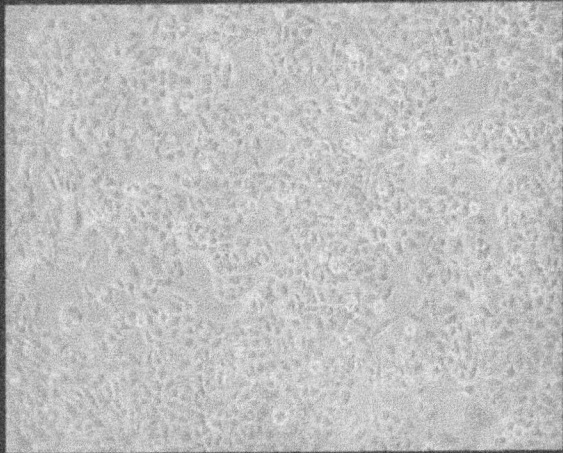
MDCK cell line—Normal (100×)

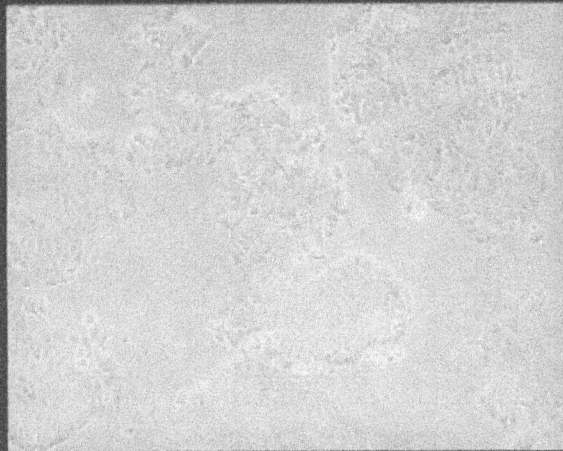
MDCK cell line—Infected showing grouping and sloughing of cells as cell sheets (100×)

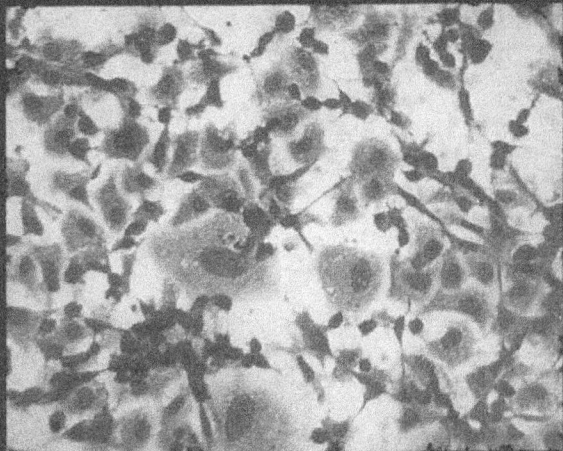
Progressing cytopathic effect showing vacuolation of cytoplasm and granulation of nucleus (100×)(H & E staining)

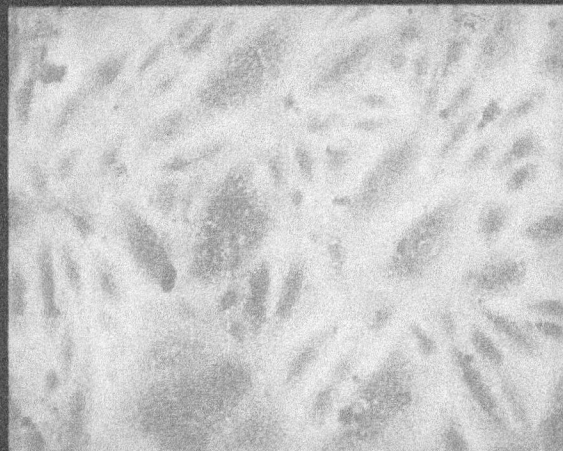
Virus-infected vero cells—syncytia, H&E (200×)

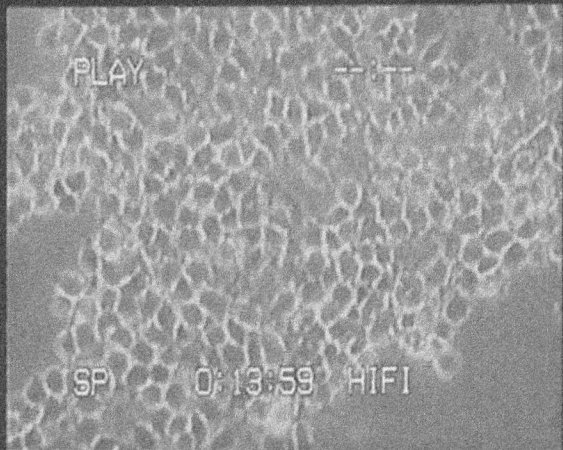
Trypsinization of the cell monolayer (200×)

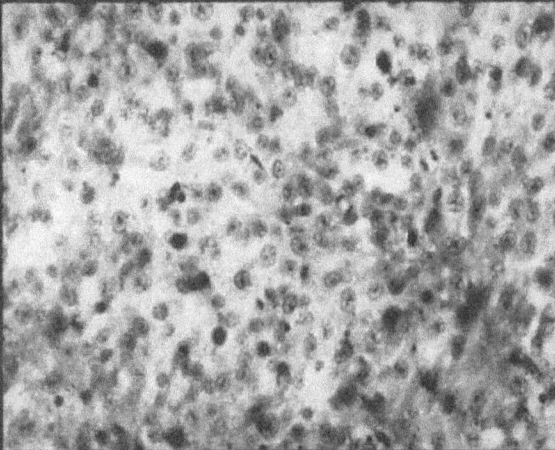
BHK-21(Razi)—Normal (40×)

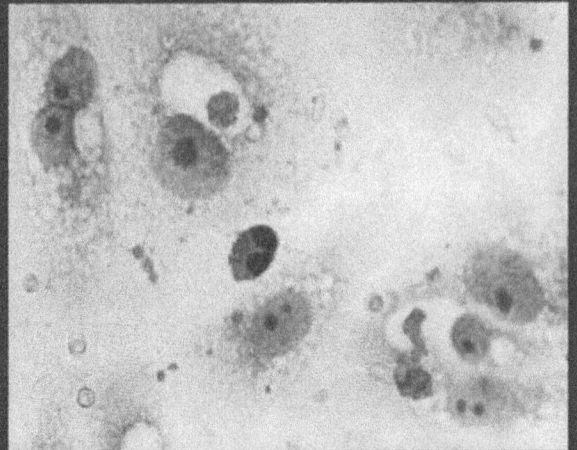
Infected BHK-21(Razi)
showing intracytoplasmic inclusions (200×)

Chicken embryo fibroblast—normal (40×)

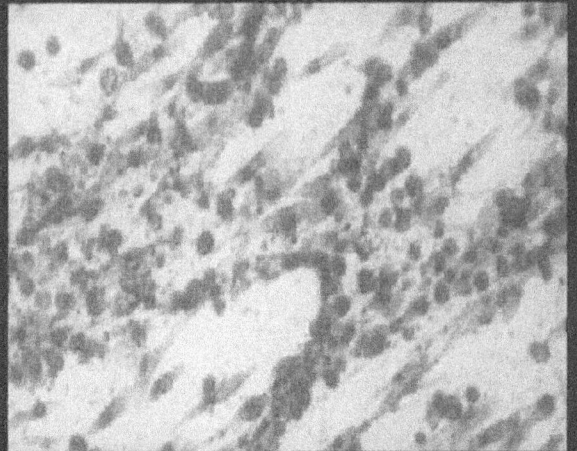
Chicken embryo fibroblast—Infected (40×)

Complete monolayer formation

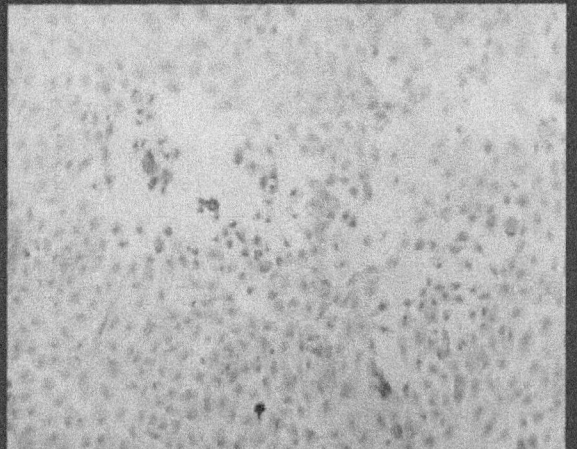
Old cell monolayer getting detached from the
attachment matrix

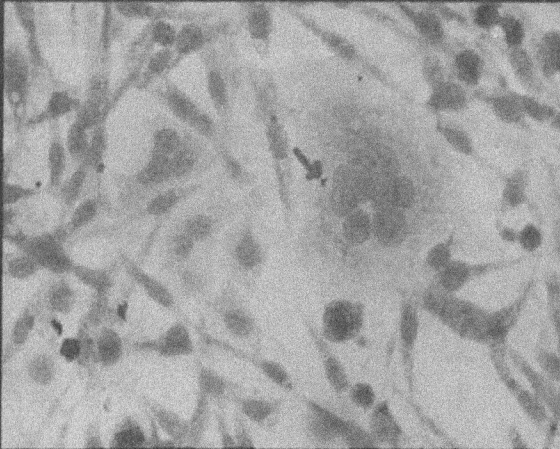
L929 Mouse fibroblast cell line showing formation of polykaryon on infection with arthropod borne viruses (200×)

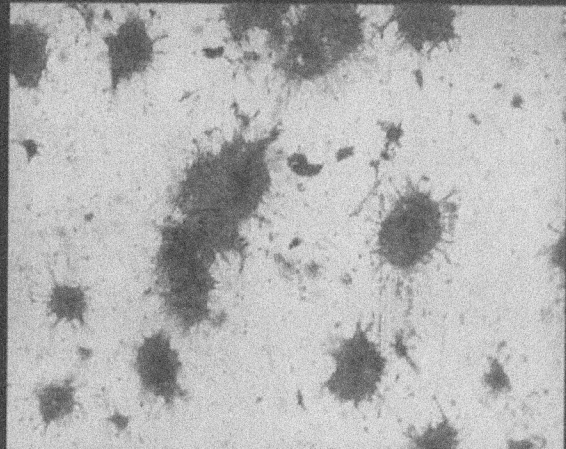
Cell line showing apoptosis (40×)

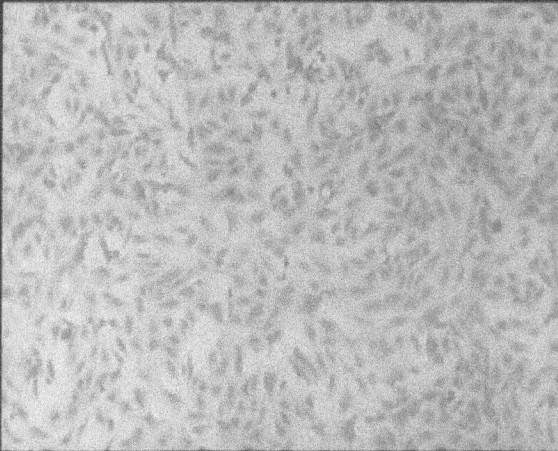
Normal cell line

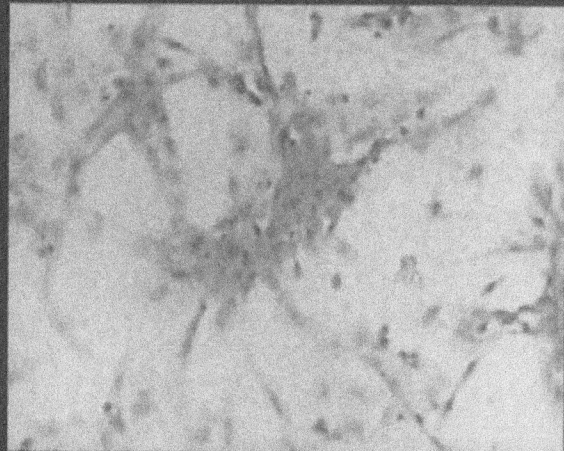
Infected cell line showing cytopathogenic effect

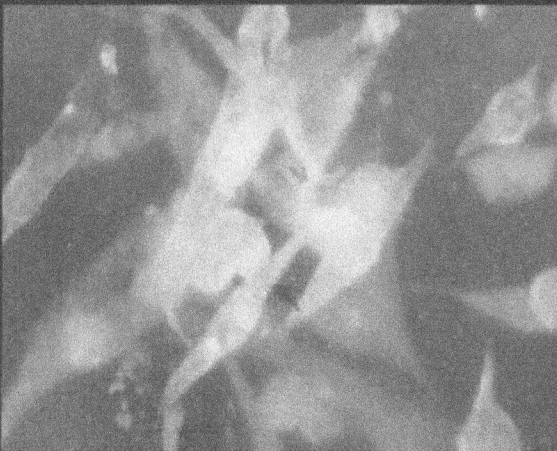
Acridine orange staining used to differentiate DNA and RNA viruses based on the orange and apple green respectively

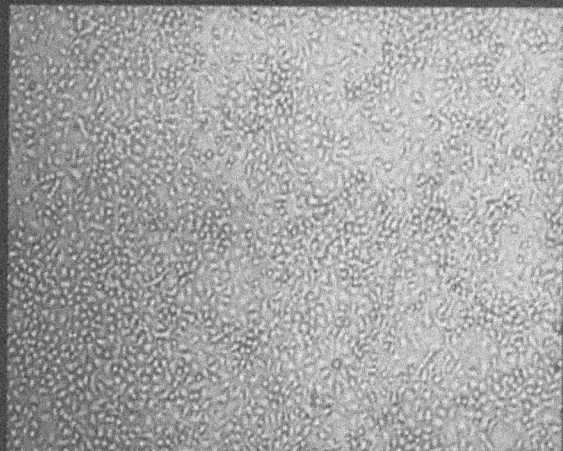
MTT assay of anti-cancerous substance in a cancerous cell line—8 hours incubation (40×)

Cytopathogenic effect in African green monkey kidney cell line—Vero

Seeded

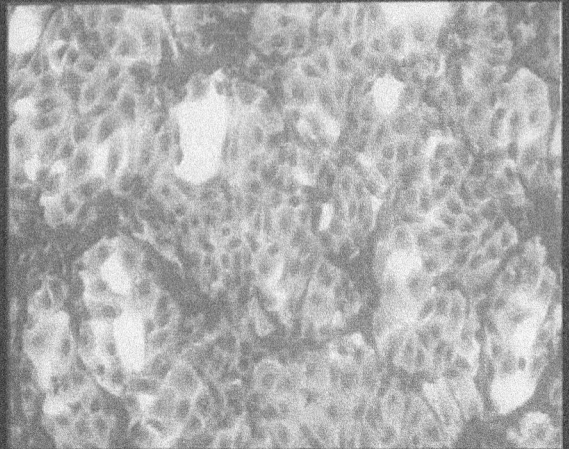

48 hours

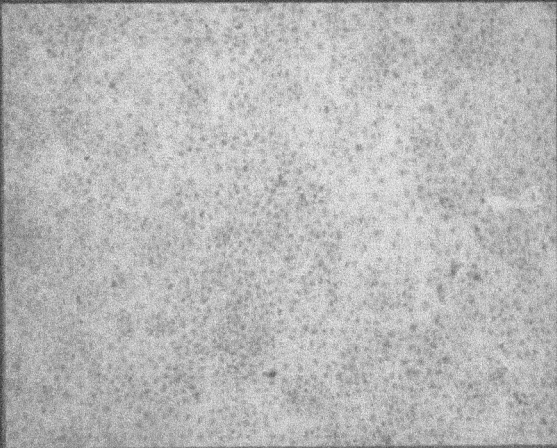

24 hours

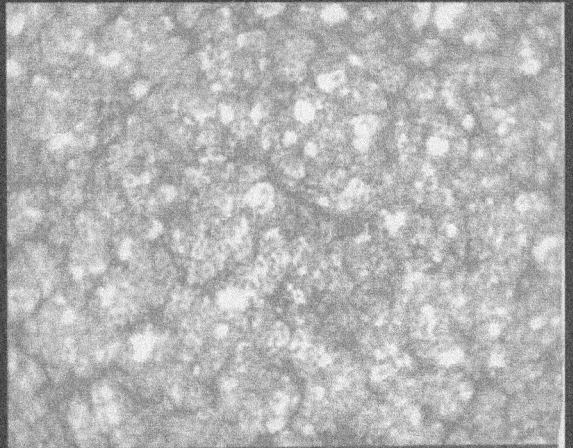

72 hours

C6/36 - *Aedes albopictus* (Insect) cell line

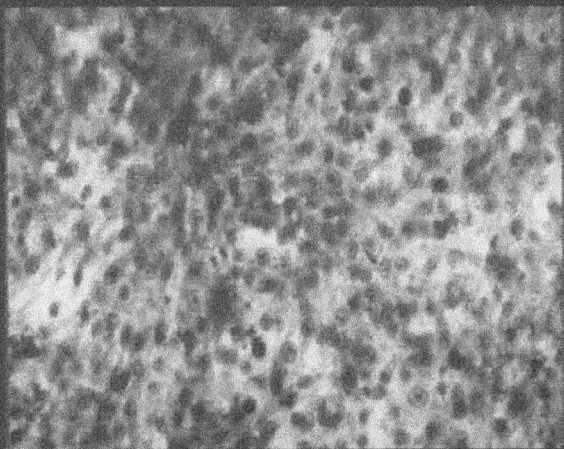

Normal

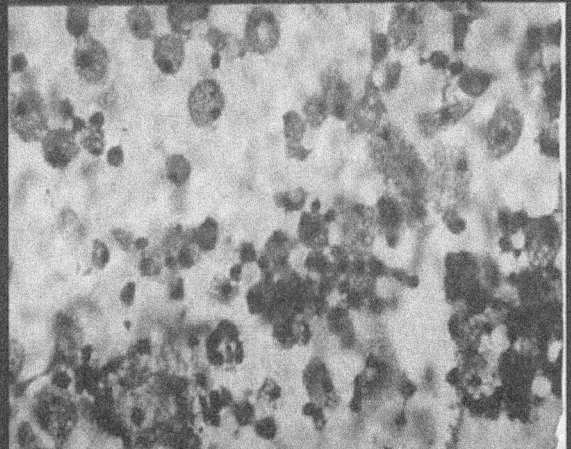

Infected with Dengue virus